D0597656

$6.95

Finnish Design: FACTS AND FANCY

Donald J. Willcox
photographs by Ove Hector

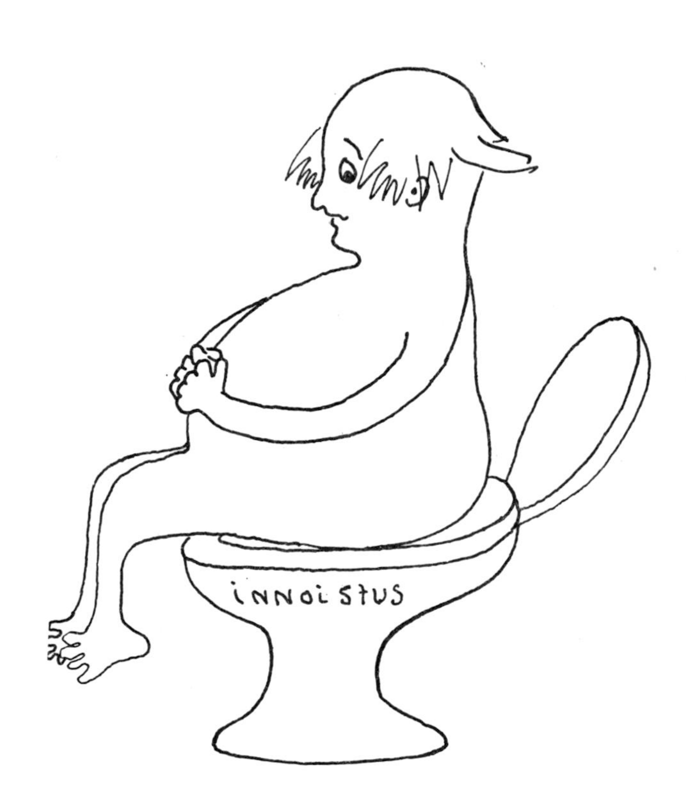
innoistus

'Truth is not the secret of a few'
yet
you would maybe think so
the way some
librarians
and cultural ambassadors and
especially museum directors
act
you'd think they had a corner
on it
the way they
walk around shaking
their high heads and
looking as if they never
went to the bath
room or anything
But I wouldn't blame them
if I were you
They say the Spiritual is best conceived
in abstract terms
and then too
walking around in museums always makes me
want to
'sit down'
I always feel so
constipated
in those
high altitudes

From Lawrence Ferlinghetti's
A Coney Island of the Mind

VNR

VAN NOSTRAND REINHOLD COMPANY
NEW YORK CINCINNATI TORONTO LONDON MELBOURNE

?

Finnish Design: FACTS AND FANCY

Donald J. Willcox
photographs by Ove Hector

The author and publisher gratefully acknowledge permission to use an excerpt from Lawrence Ferlinghetti, *A Coney Island of the Mind*, Copyright 1955 by Lawrence Ferlinghetti. Reprinted by permission of New Directions Publishing Corporation.

Van Nostrand Reinhold Company Regional Offices:
New York Cincinnati Chicago Millbrae Dallas

Van Nostrand Reinhold Company International Offices:
London Toronto Melbourne

Library of Congress Catalog Card Number: 72-12552
ISBN 0-442-29470-0

Printed in Finland by Werner Söderström Osakeyhtiö 1973

Designed by Urpo Huhtanen

Published in 1973 by Van Nostrand Reinhold Company
A Division of Litton Educational Publishing, Inc.
450 West 33rd Street, New York, N.Y. 10001

Published simultaneously in Canada by
Van Nostrand Reinhold Ltd.

16 15 14 13 12 11 10 9 8 7 6 5 4 3 2 1

About The Cover

The cover of this book shows a photograph of a Finnish stop sign. These stop signs were fairly common in all parts of Finland. This particular stop sign was located in Helsinki, about two city blocks from the Lauttasaari bridge. The photograph was taken with a Nikon camera using a wide-angle lens.

A red triangle within a white circle, within a red octagon, is the standard international road symbol for stop. The black hand has been a Finnish addition.

With what we can see from the photograph, the black hand on the stop sign is held up, palm thrust forward, with only the silhouette of a black palm extended toward us. Who does this hand belong to? Why did it appear only in Finland, and why was it Black?

If somebody were to thrust a black hand at me, I would not be able to interpret this as a friendly gesture. On the contrary, I would take it as a display of *authority*, and I would find it entirely unfriendly.

Does a Finn need authority? Does a Finn need a black hand thrust at him? Does he need to be continually reminded that he is not far removed from the forest, and that he is not to be trusted — even by his own?

I have seen this hand in many places. I have seen it extended by a uniformed guard standing in front of Finnish restaurants, factories, and even nightclubs. Frankly, the presence of this hand frightens me. I wonder if it frightens you. I think what frightens me about the black hand is that it is a leftover symbol from centuries of human and social inequity. It frightens me because it casts the Finn into a role which he outgrew and matured from generations ago. If the authority defines a role, will there not always be actors to play the role?

How many Finns do you know who need a black hand of authority? I do not know one!

Thankfully, these black hands are getting old and rusty. They seem to be having trouble holding themselves up.

Dominus vobiscum!

for my woman . . . MARITA . . . who puts cabbage leaves on my bad knee
and does
other ? things
XXXOOO

Contents

Foreword

**Note* — The following interview is a figment of your author's imagination.

MR. SUKKANEN: "Tell me how you feel about the book I am about to read".
MR. WOOLSOX: "I feel fine".
MR. SUKKANEN: "Be serious — that's not what I mean".
MR. WOOLSOX: "Well, what do you mean?"
MR. SUKKANEN: "I mean how do YOU feel about it?"
MR. WOOLSOX: "Well, sometimes I feel that an author is a lot like the hero in a Hollywood western".
MR. SUKKANEN: "Hold on now. You can't possibly be serious. You must be pulling my leg".
MR. WOOLSOX: "Not at all".
MR. SUKKANEN: "Well isn't it a bit unfair to begin an intellectual non-fiction book by blatantly trying to compare your position to that of the hero in a Hollywood western?"
MR. WOOLSOX: "Not at all".
MR. SUKKANEN: "I'm afraid you will have to be more specific".
MR. WOOLSOX: "Well, take the case of John Wayne".
MR. SUKKANEN: "What about John Wayne?"
MR. WOOLSOX: "Okay, so John Wayne is the perpetual lonely cowboy with the purity of a priest. Right?"
MR. SUKKANEN: "Right".
MR. WOOLSOX: "He gallops through western after western on his oversized horse in the name of good and committed to the conquest of evil. Right?"
MR. SUKKANEN: "Right".
MR. WOOLSOX: "The structure of his films are built sequence upon sequence always growing toward a dramatic, nail-biting finale. Right?"
MR. SUKKANEN: "Right again".
MR. WOOLSOX: "So then, in one dramatic climax, our hero single-handedly rounds up the bad guys, and then races off to overtake the train in order to rescue the maiden who has been unfortunately tied to the railroad tracks — a victim of bad guy foul play".
MR. SUKKANEN: "Now we're getting somewhere".
MR. WOOLSOX: "Funny thing — John Wayne always succeeds. At the final breathless moment, he outdistances the train, leans over in the saddle, whips out his knife, cuts the maiden loose, saves her from extinction, deposits her in the arms of her joyful mother (untouched), and then rides off into the sunset — again alone".
MR. SUKKANEN: "But what is your point?"
MR. WOOLSOX: "Please try to understand. What the movie viewer expects to see is PERFECTION. He has been taught to expect PERFECTION — and in the case of a Hollywood western, a kind of super-human perfection".
MR. SUKKANEN: "Well what is wrong with that?"
MR. WOOLSOX: "This kind of perfection just doesn't exist. It's a myth. And as a matter of fact, it's an ugly myth".
MR. SUKKANEN: "What do yo mean it's a myth?"
MR. WOOLSOX: "Hollywood has manufactured the hero and has constructed him into the living symbol of perfection. John Wayne rescues the maiden, and then captures the bad guys — all as a tightly organized consequence of everything that has happened before in the film".
MR. SUKKANEN: "I still don't see your point".
MR. WOOLSOX: "Have you ever seen John Wayne fall off his horse?"
MR. SUKKANEN: "No, I can't say that I have".
MR. WOOLSOX: "That's my point. Nobody has. Nobody ever sees the fifty rehearsals where John Wayne did fall off his horse, where he dropped his knife, lost his script, or swore at his director. And what's more, nobody ever sees the dozens of busy people on all sides of the camera who comb John Wayne's hair, brush his boots, remind him of his lines, tell him when to go to the bathroom, and encourage him to act like the super hero he is supposed to be".
MR. SUKKANEN: "Well, what has all this got to do with your statement about feeling like John Wayne?"
MR. WOOLSOX: "I didn't say I felt like John Wayne".
MR. SUKKANEN: "Well what did you say?"
MR. WOOLSOX: "I said I feel that an author is a lot like the hero in a Hollywood western".
MR. SUKKANEN: "How so?"
MR. WOOLSOX: "Okay, to begin with, a non-fiction book is supposed to be a kind of holy exercise of perfectly organized, and hygienically clear fact statements — always building — one fact building upon another — headed straight as an arrow — toward a final intellectual orgasm where all that has gone before suddenly explodes in logical sequence to reveal new wisdom. Right?"

MR. SUKKANEN: "Well we do expect the author to be well organized, and to build his case into a clear climax".

MR. WOOLSOX: "That's it! The author's position has been defined even before he sits down to write his first word. He is expected to perform as the flawless creator who guides his reader by the hand, step-by-step, until at last the cymbals crash, the heavens roar, and WHAM . . . NIRVANA — the point of the book is revealed in all of its logical splendor".

MR. SUKKANEN: "This is beginning to sound just like a Hollywood western".

MR. WOOLSOX: "That's exactly it! Now you're getting it! The very medium through which the author hopes to express himself has already laid down a procedural structure of rules called a book, and the author is expected to perform precisely like the perfect machine that he is not, and never will be".

MR. SUKKANEN: "What do you mean by rules? I thought writing was one of the highest forms of free expression".

MR. WOOLSOX: "Whew . . . it's hot in here! The only reason I call them rules is because I don't know what else to call them. Before the author can contact his reader, he must have the support of his editor, his publisher, the bookseller, the critic, and many other people. His work is seen and indirectly controlled by these people, and when you get a lot of people working on the same thing — plus a whole publishing tradition — you get a kind of organization or structure started. And then you get these kinds of rules of the trade like operational procedures".

MR. SUKKANEN: "What kind of rules?"

MR. WOOLSOX: "Well, for example, rule number one is that the author is expected to be the symbol of intellectual perfection. He is given two covers, and between these two covers he is expected to always be logical, and of course to build his book within a tight, clear structure".

MR. SUKKANEN: "Still sounds to me like a Hollywood western".

MR. WOOLSOX: "Or maybe John Wayne? Can you see the similarity now? John Wayne the symbol of masculine perfection — versus — the book author as the symbol of intellectual perfection".

MR. SUKKANEN: "Go on, I want to hear more".

MR. WOOLSOX: "Well, rule number two is that an author is expected to arrive in your hands as an immaculate conception".

MR. SUKKANEN: "What do you mean by that?"

MR. WOOLSOX: "Well, consider, for example, what you and I are doing right now. We are having an informal discussion in my own home over a cup of coffee. Under these conditions I can be completely at ease. I can talk to you just like anyone else, and you can go home from here thinking that I'm quite an ordinary, even likeable guy. Now contrast this to the way I am supposed to behave when I'm speaking to you in a book. I'm supposed to be formal, I'm supposed to be brilliant, and as a matter of fact, I'm supposed to be exactly what I cannot be because of my medium".

MR. SUKKANEN: "Just like John Wayne?"

MR. WOOLSOX: "Exactly! As the reader of my book, you are never allowed to see my crumpled pages of mistakes, my poor spelling, my hundreds of bad starts, my ash-tray overflowing with cigarette butts, my unshaven face, my ragged shirt collar, or the liters of coffee I have consumed in order to come to you as the symbol of perfection. You are not allowed to share in my struggles, see me walking around the room, watch me jump up out of bed to scribble a note, follow me to the store as I waste nervous energy, and you are never allowed to sense even a trace of my unsureness, or witness the tears of pain or frustration I have spilled upon my manuscript pages in order to arrive in your hands as an immaculate conception".

MR. SUKKANEN: "Do you actually do all these things?"

MR. WOOLSOX: "Of course I do — most authors do, and we do this in order to live up to the image we know is a myth".

MR. SUKKANEN: "How else does it affect you?"

MR. WOOLSOX: "Well, for another thing, it affects my language, I mean my book language. My mind has one language, and this is fine for informal coffee conversations like this, but it breaks the rules for book language. And so, in order to keep to the rules, my mind language is set down on paper, and then it is examined, and re-examined. It is finally translated into a much more sophisticated book language".

Mr. Sukkanen: "Can you give me an example?"

Mr. Woolsox: "Well, what happens with a lot of authors is something like this. They know it is cold out, but they can't say this in a book because it sounds too banal according to the rules. And so they deliberately translate "it is cold out" until it sounds something like this — "after careful consideration of high pressure areas, divided by low pressure areas, it is reasonable to assume that we are experiencing something of the latter".

Mr. Sukkanen: "Did you just say it was cold out?"

Mr. Woolsox: "Sounded pretty fancy didn't it?"

Mr. Sukkanen: "Wow! Do authors really do this?"

Mr. Woolsox: "Of course we do. How else do you think we get so intellectual? The structure of a book gets us pondering over words, gets us word conscious, and what it finally does is to destroy the kind of immediate spontaneity that you and I are sharing just now over this coffee".

Mr. Sukkanen: "But this kind of spontaneity is the way life really is, isn't it?"

Mr. Woolsox: "Life is certainly no Hollywood western".

Mr. Sukkanen: "Does it do anything else besides making you word conscious and less spontaneous?"

Mr. Woolsox: "Yes, perhaps the final rule for an author is that it is grammatically incorrect for him to be fragmentary".

Mr. Sukkanen: "But isn't life made up of fragments?"

Mr. Woolsox: "Life is certainly no Hollywood western".

Mr. Sukkanen: "I mean, for heaven's sake, let's be reasonable. We all think in disorganized fragments don't we? Our visual images are filled with fragments aren't they? Our sound images come in fragments don't they? I mean, if all of life is made up of fragments, then why can't an author PRESENT HIS BOOK THOUGHTS IN FRAGMENTS?"

Mr. Woolsox: "Careful, I think you're going to spill your coffee".

Mr. Sukkanen: "WELL, THIS IS MAKING ME ANGRY".

Mr. Woolsox: "Ooops! Here, you can borrow my napkin".

Mr. Sukkanen: "But I mean, I've never thought about the author's position before. He's a lot like the hero in a Hollywood western. I mean, what are YOU going to do about all this in the book?"

Mr. Woolsox: "Just try to be honest. And human".

Mr. Sukkanen: "But will your publisher allow this? CAN A DESIGN AUTHOR BE HUMAN?"

Mr. Woolsox: "Excuse me, I'll have to run now. The movie begins in a few minutes. I don't want to miss a single second. Haven't seen John Wayne in years".

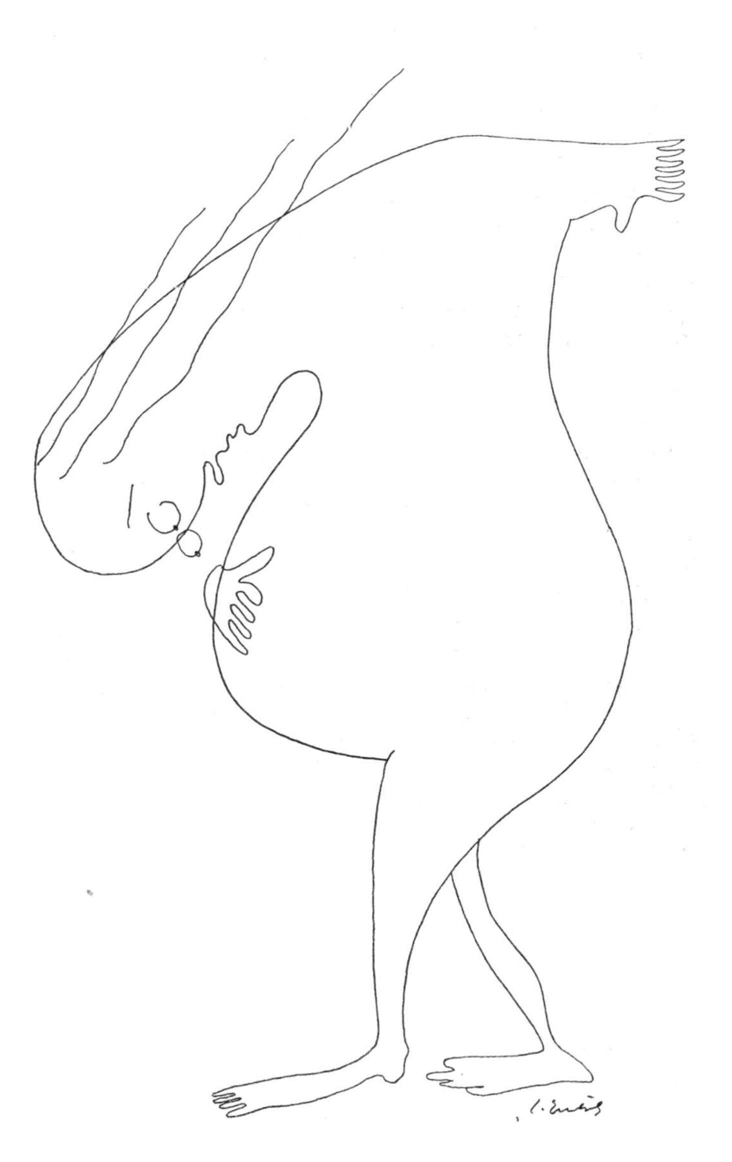

Introduction

The Parasites

The production of objects affects the lives of three distinct groups of people. First there are those who fashion and produce the objects. In general terms, one can call these people the *doers*. Next, after the objects have been produced and marketed, there is a vast audience who must either consume or reject them. One can call these people the *users*. Finally, way over on one side, and very often masked in ambiguity, there is the third group — the *parasites*. The *parasites* include that group of people who have set themselves up in a special position to live off of the energy of the *doer*. The *parasites* include the professional critics, the academicians, and the paid reviewers. They have established what they consider to be an independent "art form" which, for all practical purposes, lives off of the work of the *doer*. The *parasite* believes he is performing a valuable function midway between the *doer* and the *user* — he feels he is a testing device for evaluating the performance of the *doer*, while at the same time protecting the interests of the *user*. I am a *parasite*.

Curiously enough, we *parasites* come in several varieties. Some of us have established our own special branch of *parasitism* which is called the *aesthetic* branch. This branch has a very large membership. We have followed the lead of the social scientist and have developed our own exclusive language. This language lies somewhere between Guruu and ?#%&+?. Mostly, it is incomprehensible even to us, but we have tricked you into thinking it is brilliant. We can rattle off sentences which are so superior that we can actually make you feel that you should wash your hands with strong soap before reading us. The *aesthetic* branch of *parasites* concerns itself with concepts of *good* and *bad*. How, you ask, do *parasites* always know what is *good* and what is *bad?*

Well, I really should not give away professional secrets, but since you are curious, I'll do it anyway. Now please bear in mind that there is a lot of magic involved in the process. The *aesthetic parasites* gather together once a year for an annual meeting. This meeting is a secret ceremony half way between an American Indian Pow-Wow, and a Finnish Tango.

Aesthetic parasites gather from all parts of the world. Each *parasite* brings a package of spaghetti noodles with him. Along the edge of each noodle, he writes the name of something he intends to review during the next twelve months in his country. Now the *aesthetic parasites* all form into a huge circle around two enormous iron cooking kettles. A fire is built under one kettle, and then water is poured in. This kettle is labelled *good*. The other empty kettle is labelled *bad*. As soon as the water boils, each *parasite* dances past kettle *good*, and drops in his package of spaghetti noodles with the names printed on the side of each noodle. The spaghetti is allowed to boil for exactly twelve minutes with a pinch of salt added. The noodles are then stirred vigorously by the Chief *parasite* while he hops around the kettle singing aesthetic songs and dressed only in his underwear. When the stirring is stopped, several subchiefs pour half of the full kettle into the empty kettle labelled *bad*. As soon as the mixture has been allowed to cool, the spaghetti noodles are given back to their rightful owners in two separate piles — one marked *good*, and the other marked *bad*. The distribution of the noodles from one kettle to the other is entirely a miracle. Many *parasites* feel the miracle has something to do with the number of rings around a cat's eye.

Now that the *aesthetic parasites* have been given this miraculous information, they can return home and pass along this final judgment of *good* and *bad*. And they do.

The second branch of *parasites* was formed as a result of a schism with the *aesthetic* branch. This alternative branch has rejected the concept of *good* and *bad*. They also get stomach cramps when they hear the words "art" and "design". They call themselves the *common-sense* branch, and actually, they are quite dull people — most of them are former working people — even truck drivers. I am a member of the *common-sense* branch of *parasites*. We have very few members. No Sherry is allowed at our annual meetings. We mostly drink milk. In preparing our critiques, we have only five rules: Is it necessary? Does it work? Can it work better? Is it quality? Is it reasonable?

Chapter 1

DeeeeeeeSign

No matter how hard I try to prevent it, this book will end up being labelled a "DeeeeeeeSign" book, or more specifically, a "DeeeeeeeSign" criticism.

What Is Design? Well, to begin with, the word "design" is perhaps the most commercially abused, and violated six letter word in language. The word "design" shares a linguistic position along with the word "freedom" as one of the mightiest whore words of the twentieth century. There is no longer any doubt that the word "design" is a prostitute. It has been sold on the commercial marketplace, passed from hand to hand, and has been hammered, re-formed, and stretched to aesthetically justify everything from dehydrated cow manure — to disposable diapers. As a verb, it has been prostituted as an artistic substitute for everything from sewing (designing) a dress, to formulating (designing) a political theory. As a noun, it has been misused to identify everything from soil conservation to birth control. And although theoretically, it is an English word, it has been internationally prostituted to account for everything from *A*fghanistan coats — to *Z*en Buddhist yoga.

What Is Design? Above all, the word "design" has become irretrievably entangled with the words "profit" and "money". For the man with something to sell, "design" means an aesthetic stamp of approval — an identity device — which implies an object of excellence. It has become entirely too easy to replace craftsmanship, quality, and function by simply calling an object "design". With this new, unearned label, the object not only sells, but it sells at a higher profit.

What Is Design? In its contemporary context, the word "design" either means everything or it means nothing. I believe it has been slaughtered into so many inconsistent, and diametrically opposed parts, that it now means NOTHING. Because it has been linguistically abused in so many ways, it is no longer a reliable term for an exchange of intellectual communication.

What Is Design? If one is to believe what critics and designers say about the subject of design — then the confusion over this single word grows even heavier. On the one hand, there are those who preach a doctrine of aesthetics — with the added postscript — to hell with people. On the other hand, there are those who shout — to hell with aesthetics, and LONG LIVE PEOPLE. For whatever design is, or is not, the point of expressing design energy seems to lie somewhere in between these two extremes — perhaps just a bit left of center — somewhere at the point of aesthetics with a soul, or better yet, refined form with a conscience.

Is this therefore a book on design? Curiously enough, it will be called just that, even though I wish there were some way to consistently call it a book about objects. In no matter which direction I attempt to break away, I end up being stuck with the word "design" whether I like it or not. So far, our language just hasn't come up with a reliable substitute. Whenever possible, however, I will try to keep away from the misuse of "design" by referring to forms as objects. Even though this word "object" is not as romantic or mysterious as the word "design", it is at least a bit more reliable, and less emotionally charged for a communication of this kind.

Specifically, my task herein is to explore the objects of Finland — the background and environment out of which the objects of Finland are produced. The word "object" is extremely broad, and allows freedom to explore form in every direction. Whereas the word "design" (used in the normal context of aesthetics) — limits us to beautiful forms, the word "object" — allows us to explore at will.

Over the past 40 years, and like many so-called western nations, Finland has expended a staggering sum of individual, cultural, and economic energy toward producing objects. One can now honestly observe, free of political overtones, that Finland is in fact, a highly object-oriented nation. The number of separate objects now produced in Finland number into the tens of thousands. Except for certain imported raw materials and a number of imported foodstuffs, Finland is not only self-sufficient in object production, but is an aggressive competitor on the international export exchange.

In return for this expenditure of object-producing energy, Finland has gained an international reputation of excellence, particularly in so-called "aesthetic" objects for the home. In many circles, especially elsewhere in Scandinavia, this reputation of excellence often inflates to point of romanticism — one now commonly reads and hears the expression "exotic Finnish flair".

Several books, dozens of magazines, at least a half-dozen films, and dozens of exhibitions have been fully committed to spreading this "gospel" of Finnish excellence with its "exotic flair". The literature directed to this end is in great abundance — it is given out in travel offices, airline offices, press bureaus, trade centers, embassy offices, and anywhere else where Finnish interest is at heart — and it is handed out absolutely free.

Marimekko textiles have made the pages of "Life" magazine, Futuro houses by Polykem have made the pages of "Playboy" magazine, and in the fall of 1970, the whole Copenhagen walking street was bowing to the excellence of Finnish object production. Ulf Hard af Segerstad has devoted a whole book to "Modern Finnish Design"; the Praeger Publishing Corporation in New York has produced a special series of books devoted exclusively to the various "arts" of Finland; the American design critic Victor Papaneck has carried the banner of Finland into international visual education; the Finnish Foreign Trade Association annually publishes 74 pages of "exotic Finnish flair", and perhaps the most spectacular presentation to date, was the December, 1969 issue of "Avotakka" magazine done in a rainbow of colors with a Finnish, Swedish, English, German, and French text (Swahili was somehow omitted).

At this point, and in view of the already solid reputation of Finnish excellence, one might well ask why still another book about Finnish objects is really necessary? If a maximum reputation has already been gained, and if this reputation now rests securely locked in the vaults of commercialism, why then is it necessary for still another pat on the Finnish back?

PAUSE. Coffee break!

(Now comes the moment of TRUTH! So let the trumpets blare, and let the curtains open wide. Your author is about to be stripped to his bare skin and revealed for what he *really* is. He is about to be cast in his true role—THE DEVIL.)

END OF PAUSE. End of coffee break!

THIS BOOK IS NOT INTENDED AS ANOTHER PAT ON FINLAND'S BACK!

Through the pages that follow, my hand has slipped lower down along the sensitive Finnish spine and is reaching out to give Finland a pat gently on her BUTTOCKS!

Now one must understand that a pat on the buttocks can be administered in several ways. For example, one way is to attack viciously and with explosive anger. Another way is to tiptoe into position for a sneak attack. A third approach is to play the authority role of the father figure spanking his ill-behaved child, and still a final approach is to direct the pat as a gentle form of positive encouragement.

Those of us who have managed to survive into the 1970's are abundantly aware that the 1970's have become a time of protest—a time of sit-ins, love-ins, be-ins, rock-ins, smoke-ins, walk-outs, and march-outs — where hardly a single institution or tradition is exempt from attack. Too much of the energy involved in these attacks is negative — everyone seems to be *against* the "establishment", *against* tradition, or *against* the institutions we have inherited. Too few have become POSITIVE voices of change — voices that attempt to bring about a general improvement, or voices which provide alternatives.

It is not enough to simply join a battle-cry of revolution. Revolution is aimed as an end to whatever exists. When something is ended, the result is a ZERO. Many voices feel that a ZERO is the only place where change can begin. But change IS NOT A SYNONYM FOR IMPROVEMENT. Revolution may well cause change, but REVOLUTION DOES NOT GUARANTEE IMPROVEMENT. Improvement begins with self-examination — the willingness to look oneself in the face and evaluate an expenditure of energy, and the means and direction of progress at which this expenditure is aimed.

This, then, is the task of this book—to provide an open forum, an accounts ledger, or a kind of mirror wherein Finland can examine herself in the production of objects. Yes, it is true that this book is a criticism, but it is not a criticism aimed at producing ZERO. It is a positive criticism aimed at encouraging alternatives and improvements.

Now at this point, my reader may well assume that some of my cups are not in my cupboard. I have revealed the task of the book without justifying a need. On the one hand, I admit that Finnish objects enjoy a reputation of excellence, but on the other hand, I propose a book that criticizes Finland with a gentle pat on her buttocks. How can a criticism be in order if a reputation of excellence is already secure?

Well, to begin with, perhaps we should take a hard look at this word "reputation". Now, no single star in the human firmament ever glistens brightly without a well-organized, and massive advertising campaign. This seems to be the first rule in promoting everything — from a tube of toothpaste — to a can of spray deodorant — to a Hollywood sex symbol — yes, to even an object-producing nation. If you plan to sell — then it pays to advertise. If you hope to sell well — then it is profitable to build an image — a reputation.

It will come as no surprise to anyone that Finland, as one of the more aggressive object-producing nations, has engaged in a massive internal and external advertising campaign. This program of reputation building — from the inside to the outside — has cost millions of Finnmarks, but it has brought about effective results — the reputation appears at least temporarily secure. But advertising, which is a means toward a profitable end, is not a synonym for honesty. It is therefore altogether healthy and proper that advertising and reputation building be openly evaluated in order to secure honesty. In other words, how does the reputation of Finnish object production, gained by advertising, compare with not only the objects, but with the environment out of which those objects are produced?

Advertising, whether for a specific cosmetic or for an entire object culture, can be, and is very often disguised under many other names. To disguise advertising so that it no longer looks like advertising is to catch the public unaware. Some of the more common masks to disguise advertising include — presenting it as a public notice, a consumer report, a slice of goodwill, or worse, as a chunk of "education". What happens, for example, is exactly the same as what happens on television, or in the movie theater before the film begins. One sees a middle-aged, goodlooking man dressed up in white to look like a doctor. He tells us that brand "XYZ" toothpaste has been scientifically tested effective in the prevention of bad breath. We are supposed to believe him. His white doctor's uniform is offered as our proof. When you and I see this on television, we don't take it seriously because most of us know it is nothing more than a lot of ground-up verbal hamburger. Other examples are not so easy to spot, and in fact, the concept of advertising is so subtle that whole masses of people can be spellbound without ever suspecting it.

Let me here give you two vivid examples of how self-advertised, cultural reputation building can be carried to shocking extremes. As a child, and then as a young man growing up in America, I spent 18 years as a student in the American educational system. This formal education carried me from kindergarten through grade school, and from high school into advanced study at three different American universities. Like most inquisitive students, I not only read my required school books, but I did a fair amount of additional reading.

I had been educated to believe that George Washington was the "father" of his country. His birthday is celebrated each year with a National holiday, and his life and work are held as the ultimate human achievement in decency, honesty, and the securing of individual rights. George Washington was, and still is, the symbol of perfection at which every "good" American boy should aim his life. Not once, during all those years of "education", did I ever read an unbiased evaluation of George Washington free from extreme romantic prejudice. Although the beloved "father" of his country owned and used some four or five dozen black human beings like they were pigs and cows, this fact was conveniently omitted from American schoolbook evaluation. This action, even today in most American schools, is exempt from study.

And like all young boys growing up in America, I was an ardent follower of the historic colonial migration from the east coast to the west. Like many million other school children, I played cowboys and Indians, read saga after saga in my schoolbooks, and saw dozens upon dozens of films where the "patriotic" white man had triumphed over the "savage and brutal" Indian. I was taught to identify with the courageous settler and to consider his actions as honorable and just. Not once, during all those years of "education", did I ever read in a book that the extermination of several hundred thousand American Indians could be other than romantic and patriotic — another slice of American history conveniently omitted from schoolbooks.

Had these two extreme examples not happened, the average rational man would assume that when in community with other rational men, something like this simply could not happen. But most intelligent Americans, in fact, do consider themselves rational — and yet, these two historical omissions persist even in the 1970's (my own children, one generation later, have been kept from these same two facts in their own schools). I record these two examples (and I could record many more), only to point out that the very foundation on which a whole national self-image is built, can in fact result from irresponsible self-advertising — or, as with the two examples illustrated — the historical face-saving for a population of over 200 million people.

Admitting, then, that there are dangers within self-advertising — dangers which seem to be inevitably built into the ego medium — how, then, can this be related to an evaluation of Finnish object production and its admitted reputation of excellence?

Honesty.

Honesty, a six-letter word — easy to write, simple in theory, but yet extremely painful in practice. If the reason for most of the anguish and challenge of the 1970's could be summed up into one word — that word would be HONESTY. The cry around the word of "tell me where it's at" is a cry for honesty — it is a youthful voice pleading with his adult world for a return to honesty. Honesty *is* "what it's all about", and it is the key, the pulse, and the conscience wherein the solution lies for most of the problems confronted by this book.

To begin to approach honesty, one must propose many questions and then be willing to supply straight answers — questions and answers which do more than spin the normal clouds of policital abstraction. Abstraction is a particular gift of the politician — honesty, on the other hand, is not one of his strong points. I am not a politician. Therefore ambiguity is out of order in a communication of this kind. No matter how hard it hurts the ego or the pocketbook, it is altogether healthy for Finland to take an honest look at itself in the business of making objects.

I can already hear one hundred loud voices — intellectual voices much wiser than my own — raised in complaint that honesty cannot be defined, that honesty is an unworkable concept, that honesty is dead, that there is no such thing as honesty, that honesty to one man is dishonesty to another — and on they go, one voice shouting after another.

What then is honesty? Can anything be honest?

Honestly, I cannot define honesty, but I believe I can begin to *apply* honesty. Honesty, no matter how slippery it might be to define, has something very intimately to do with our primitive instinct of common sense. Yes, we have lost our ability to reliably define honesty, but we have not lost our instinctive ability to *apply* honesty through common sense.

Common sense.

Take a look at it. Common sense looks like an antique sitting on the page. Our contemporary intellect has become so super-sophisticated that common sense sounds about as ridiculous as trying to light a sauna fire by rubbing two wet sticks together in a windstorm.

Yes, for the contemporary thinker, the concept of honesty may well be suffering from incurable cancer, but our contact with our own instinctive common sense still lingers on — at least some of mine has, and I hope some of yours has as well. If, for example, you were to set a coffee cup in front of me, I can begin to evaluate this cup with none other than my common sense. I can begin to discover how honest this cup is by simply putting this cup to the test. Does it fill a need? Is it made of quality material? Is it well balanced? Does the handle fit my finger? Is the drinking edge smooth? Does the material burn my finger? Does the material insulate heat or conduct heat? Is the form convenient to stack and to wash? Is the cup reasonably priced? Does its form begin to harmonize with its function? And on it goes. I can use this cup, experience it in my life and its life, and then report nothing but the results of my common sense — an attempt to *apply* honesty.

I hope I am beginning to make sense — common sense? If I exercise my common sense, if you exercise your common sense, and if we together exercise our mutual common sense and apply it against the needs and problems of our environment — we will begin to apply the nearest living relative to honesty.

We can apply our common sense against all aspects of Finnish object production, and we can begin to evaluate what has happened and where it is headed. We can apply a common sense evaluation against architecture, against household objects, against other products, against planning, against social values, against advertising, against education, against the environment in which the designer must function — and we can together gain a valuable perspective on Finland with which to formulate improvements. We can ask ourselves otherwise embarrassing questions, we can begin to supply non-ambiguous answers, and we can explore alternatives. Through this medium of grossly underrated common sense, we can begin to focus sharply upon our own *real* image as opposed to our *fantasy* image.

How is it that an American is writing a book on Finland?

In 1966 I was at an American university in Vermont, and at that time I was writing a very technical book on handwork. For perhaps ten years prior to 1966, I had been following the promotional image of Finland — at least that part of the image that arrived in my hands in the form of promotional magazines, brochures, and occasional articles in American publications. The message of excellence admittedly stimulated my curiosity, and whenever possible, I would drive down to Boston, Massachusetts, to visit the Design Research shop in Cambridge to see for myself the imports from Finland.

All of us go through various stages in our lives, and I was then in my antiseptic "aesthetic" stage. I ended up filling my house with Knoll Associates furniture by Saarinen, and adding a few sprinkles of other Finnish products around my home. I was of course very pleased with myself and with my good taste in selecting only "exotic" Finnish flair. I was then not only a distant admirer of the Finnish flair, but had done something positive about it by surrounding myself with frightfully expensive, but *real* Finnish design.

Just about that time, or soon after, I became acquainted with a Finnish couple teaching at the university on a one-year visit. My curiosity about Finland was growing into an obsession, and I was forever asking my friends for more information to read and digest. There had not yet been a book published about Finnish objects, and I was forever complaining because information was available in only such small bits and pieces. Perhaps in jest, or perhaps because they were irritated with my imposing hunger for information — my friends suggested that since I was already an author, and since there was thus far no book on the subject — I should get up off my middleage, go to Finland, and write my own book.

Little by little this seed of an idea began to take form, but this was still in 1966 with many thousands of kilometers yet to travel and many obstacles still to overcome. Perhaps the initia stimulation to pursue the idea had something to do with the common instinct shared by all men to discover "greener pastures".

As an American, I was born with an inheritance — I had inherited the "American dream" that mankind was endowed with the creativity to build a rational world. I was fed and nourished with this dream all through my childhood, and like millions of other young Americans, I believed what I was told. But then, as inevitably happens, and as my generation reached manhood and was free to explore the validity of its dream, we discovered that the dream was in direct conflict with itself. We saw with our own eyes that the dream was only a thin, fragile veneer, and that the whole structure supporting the veneer was in a tragic state of internal decay. We saw that

America — "the super nation" was America, the psychotic nation. We tried to fit ourselves into our society, but we didn't fit because we were round plugs trying to fit ourselves into square holes.

Through some stroke of fate, ill-fortune, or luck — I found myself as an author in the field of visual education. My mistake was that I took it seriously — I was in a visual field looking out at visual chaos. When caught in this dilemma, a man has but three choices. He can sweep the chaos up into neat "aesthetic" piles, he can call for an end to chaos,or he can begin to look elsewhere for the inheritance of a rational world. I tried the first two alternatives, but they were not enough — and so I began listening to what was being said about the possibilities for a rational world in Scandinavia, particularly Finland. It seemed reasonable to me then, and it still seems reasonable to me now that Finland contains all of the necessary raw material with which to build a rational society.

My first attempts at testing the curiosity of American publishers to produce this book ended sadly. Oh yes, I received very positive response from the several Americans I spoke with. When they looked my paperwork ideas over, they usually commented that the book was a "fine idea", but then, as the conversation progressed, I was usually asked something intellectual like "isn't Finland one of those iron curtain nations?" After such a remark, the meetings would quickly deteriorate and conclude by the publisher saying something like: "although the book is a splendid idea and is needed, we are afraid it will not sell". Somehow I lost interest in American publishers for the book. It seemed to me that their response was itself symbolic of the psychotic nation — "yes, it was a wonderful idea, but it wouldn't sell".

The book then took a vacation, even though it still remained first on my private priority list. During the next 3 years I was lucky enough to still realize at least part of my initial idea to explore Scandinavia, but I had to realize this by writing six other book titles that *would* sell, by American standards. These books brought me deeply into Scandinavia and finally to Finland.

As of this writing, I am no stranger to Finland, and I hope I have never once so much as breathed a single tourist breath on her shores. Instead of standing in the Helsinki market looking up at the seagulls, I have tried to dig in, settle down, and listen to the pulse of what was going on around me. By now I have become a regular commuter across not only the Baltic, but across the Atlantic between Helsinki and New York. I have experienced Finland in all her seasons, and have traveled thousands upon thousands of kilometers in every direction from Helsinki. Interviews and correspondence have brought me in contact with several hundred Finnish men and women engaged in the production of objects. The point in my listing these encounters with Finland is only to indicate that now, even after finding myself a Finnish life partner and entrusting my three American daughters in Finnish hands — a lot of Finland has rubbed off into my skin.

I came to Finland to look for a book, but I found a book already waiting to be written — a book which simply needed an author. After only my first few days in Finland, it became clear to me that I had accidentally stumbled upon a timely struggle that desperately needed airing. I had simply asked myself the question — "Finnish design, is it a fact, or is it a fantasy?" And then, as I asked the same question in Finland, I found myself involved in an electrically charged struggle of people and values. The deeper I poked my foreign nose — the clearer it became that the struggle needed a mouthpiece — somebody to simply carry the feelings, opinions, and conflicts onto the printed page.

As an American in Finland, I was in a special position. I found myself not only involved in the struggle, but in a position to express the struggle. Because of my foreign nationality, I found myself with the possibility to express what hundreds of others wanted to express, but could not — and incidentally for many good reasons. For some unintelligent reason that I'm sure I will never quite understand — I have the possibility to realize what many others could do much better. I have the opportunity to open my big American super-nation mouth and realize an audience.

But it wasn't all as easy as it sounds. I do not mean to suggest that I encountered no resistance. Yes, the resistance to this book was there, and it made itself felt in many subtle ways. By the time of my arrival in Finland, the reputation of Finnish design had reached an almost sacred pinnacle next to only the flag, the sauna, and Sibelius. I was proposing to evaluate an area which many felt was forbidden territory. Needless to say that in many circles my name is black.

It was not until I finally came in contact with Olli Alho, and later with Pekka Suhonen and Dr. Anto Leikola at Werner Soderstrom, that I began to realize the publishing platform on which these pages appear. These three, as well as countless others, have directly made this book possible. Through Dr. Leikola's help, I was fortunate enough to recieve a Suomen Kulttuurirahasto grant to help underwrite the photographic expenses of the book. Now, in retrospect, and after having opened my big American mouth in the pages that follow—I hope I have accurately reported and evaluated the trust given me.

Chapter 2

What Is Finland?

What is Finland? Far too many people would try to answer this question by simply waving an atlas, or other fact book at you. They would thumb through the pages, find the alphabetical listing "F" for Finland, and then begin their answer with the brilliant observation that Finland is a country. They would continue by reading endless lists of facts on Finland's size, its location, its population, its principal cities, its climate, its language, its monetary system, its gross national product, its principal industries — etcetera, etcetera, *ad-infinitum*. But is this Finland? Is Finland only a paper list of statistics?

If all nations can be reduced to paper lists of facts — then how can any one nation be differentiated from another? Fact lists tend to make all nations sound exactly alike. Finland has already had its share of fact books — even on the subject of design. Yes, they are delightfully informative, but they don't tell you anything!

Well then, what is Finland?

Let me begin by suggesting that Finland is not SWEDEN. Finland is not the United States, Great Britain, Germany, the Soviet Union, Japan, and it is very definitely not South Africa.

Well then, what is Finland?

Now here is still another brilliant observation taken from the pages of a fact book: "Helsinki is the largest city in Finland". But what does this tell me about Helsinki — or about Finland? Well, Helsinki is not Stockholm, Copenhagen, London, Paris, Rome, Amsterdam, Tokyo, and it certainly is not, and never will be New York City.

Do I seem to be running into an identity problem?

What is Finland?

Is Finland running into an identity problem?

Let me here make a suggestion which on my part is intended free from political overtones — and just because it may be a key to many locks, I'll print my suggestion in large letters so you can read it:

FINLAND IS FINLAND

Finland *is* Finland — it sounds easy doesn't it? All you have to do is to be yourself? Keep your own identity? But how is it possible to keep an identity in the last 30 years of the twentieth century?

Before I go even one step further with this point, I have two more slogans for Finland to hang up on the walls of the sauna:

FINLAND IS NOT THE UNITED STATES THANK GOD
HELSINKI IS NOT NEW YORK CITY THANK GOD

What then is the reality of Finland? What are the conditions omitted from the fact books which give Finland a cultural identity?

Geography

To begin with, Finland is a huge chunk of land. This mass of water-spotted land is humped up against the northwestern border of the Soviet Union. Except for Finnish Lapland, all the rest of Finland is bordered by the Baltic Sea. The Sea slices Finland off from all neighboring population centers, except Leningrad. It is therefore a reality that Finland is not in direct geographical contact with other Scandinavian population centers, even though she considers herself culturally a part of Scandinavia.

Time Gap

Some would say that Finland's geographical position constitutes isolation, but this obviously depends upon how one defines the term "isolation". It is fairly accurate, however, to say that with water separating Finland from her Scandinavian neighbors, she experiences a *TIME GAP*. Now this time gap is not a time gap in the normal sense of hours in a day. It is a time gap in the sense of an intimate contact with pulse, or heartbeat of urban Europa. The time gap is particularly noticeable in such areas as fashion, and other life styles with only seasonal duration. These continually changing style phenomena often reach Finland one season later than the rest of urban Europe — witness, for example, the length of the female skirt.

Many Finns feel that the time gap is a cultural disadvantage, but whom is it really a disadvantage for? Is it a disadvantage for the Finnish businessman? No, not really, because sooner or later, if the businessman is only patient, the time gap will close — and these fashion styles and other changing phenomena will arrive in Finland anyway. And as a matter of fact, if he is smart enough, the businessman has a much better way of anticipating changes than a businessman in one of the cities of central Europe. Does the time gap affect the businessman's contacts with the rest of Europe? Again no. He still has the telephone and the telex, and he can keep abreast of all changes simply by keeping his eyes and ears open. Whom, then, does the time gap affect, if not the businessman?

Well, his name is Mister Ultra Contemporary Finland. He can be anybody — anybody from a university student — to a sportsman — to an intellectual — yes, to even a young businessman. Mister Ultra Contemporary Finland is the man who needs to identify with whatever is "new", or in the process of change. Mister Ultra Contemporary Finland cannot stand and wait for anything. The time gap of Finland embarrasses him, and he is ashamed of his cultural lag. He feels that to wait for the latest London or Paris fashions to arrive at Stockmanns, to wait for the latest protest film to arrive at Bio Bio, to wait for the latest pop record release to arrive at Fazer, or to wait for the latest issue of "Playboy", "Mad", or "Ramparts" magazine to arrive at the news stand is an unpardonable cultural sin. Mister Ultra Contemporary Finland looks upon his cultural time gap as though Finland were geographically located next to Outer Mongolia. If it is true that the time gap mostly affects only seasonal styles — which are anyway conceptually programmed for obsolescence — is it therefore an advantage for a culture to pursue what is not meant to endure?

Climate

Let us take a look at another Finnish reality. All of Finland lies above the 60th parallel — and according to the map, it lies the same distance north as Alaska and the southern tip of Greenland. This would indicate that, contrary to what many Finnish travel posters and tourist brochures would have us believe, Finland is not at all like the Canary Islands.

For 6 to 7 months of the year it can be cold in Finland — sometimes damned cold. When one travels to Helsinki on board the *Finlandia* in January, February, or March — and with brute force witnesses the boat smashing its every meter through solid ice, within a landscape of solid ice — one can then well understand that it is sometimes damned cold in Finland. Travel posters and tourist brochures will never be able to eliminate the reality of Finnish cold.

Along with Finnish cold comes the period of mid-winter darkness. Now cold weather and snow are two elements in winter which can be adjusted to and at times even enjoyed, but mid-winter darkness is a curse upon the land. There is nothing romantic about daytime darkness, especially for those who must endure it. It spreads a mid-winter lethargy upon the land and upon the people. It is a time when most Finns feel depleted of their energy — a time when they feel physically and emotionally burned out — and a time when they feel withdrawn inward, into a kind of psychic isolation. To survive these mid-winter blues with a positive head requires just plain guts — some call it "sisu". There is no other discreet way of describing the heaviness a man feels inside when he is forced, by nature, to confront his environment shrouded in black. This darkness, then, is one more of the inescapable realities of Finland.

A Schizoid Role

As if saved by nature herself from the lethargy of winter, the Finn is plucked from cold and darkness and cast into a spring and summer which are celebrations of sunlight and warmth. The change from winter to spring is total — almost as if nature herself were celebrating the death of winter and rewarding mankind during the process. This extreme contrast between months of cold, snow, and darkness — as compared with months of almost total sunlight and warmth within an abundant and unspoiled natural environment — casts the Finn into a kind of schizoid, or dual-personality life role. Mother nature plays a trick on the Finn, and casts him within an environment of extremes. He spends his winters like an ageing bear wrapped in wool, helmeted with fur, stuffed into heavy boots, and surrounded by only grey and white — he spends his summers like the perpetual virgin dressed in sunshine, and exploding with color. While a feeling of heaviness permeates his winter — a feeling of buoyancy fills his summer. This schizoid role for both the Finn and his environment is still another of the realities of Finland.

Language

Still another reality of Finland is the Finnish language. To most foreigners, especially other Scandinavians and northern Europeans, the Finnish language looks and sounds incomprehensible. Because it does not share Germanic roots with the other so-called "western" languages, there are very few foreigners who can appreciate the richness of Finnish. Sadly enough, it is no exaggeration to say that in many circles, even sometimes elsewhere right in Scandinavia, a Finnish-speaking Finn is looked upon as if he were a man from another planet. Therefore, if the Finn wishes to communicate, it is he who must always be the linguistic juggler with his pockets filled with Russian, German, Swedish, and English. Because of his own mother language — which is useless to him except in Estonia — he must therefore learn to be the aggressive one if he hopes to keep abreast of what is happening beyond his own borders. Very few neighbors will ever initiate communication by learning Finnish. The language of Finland is therefore just like the

Baltic Sea — it cuts Finland off from her neighbors, and at the same time irritates the cultural time gap even further by presenting problems in the linguistic exhange of ideas.

Bi-lingual Myth

Finland has aggressively advertised itself as a bi-lingual nation with the claim that her official attitudes are expressed in both Finnish and Swedish. I suggest that this claim is little more than a worn-out cliche, and it has nothing to do with the reality of daily life in Finland, especially the reality outside urban Helsinki. Yes, it is true that one still finds geographical place names in both Finnish and Swedish, and it is true that many urban business establishments are bi-lingual, but the trend in recent years has become singularly Finnish and away from Swedish. When, for example, I first came to Finland, it was not uncommon to find a Swedish text running under Finnish television programs. It was then almost a rule of business to find bi-lingual menus in urban restaurants, and to find at least one clerk or waitress employed in a bi-lingual capacity. But this has changed, and I have myself experienced this change — this trend toward becoming even more Finnish and less Swedish.

Perhaps the bare facts of reality are best exposed in public transportation — which can be assumed to reflect an extension of official policy. One can now ride on public transportation which stops only in a Swedish-speaking community and find that the bus driver speaks no Swedish, although all of his passengers are Swedish-speaking. The same thing is encountered on public transportation going in and out of Finland, with the exception of the airlines. If, for example, one takes a boat out of Finland, he will find that almost everyone employed on the boat, excluding the ship officers, speaks exclusively Finnish.

These observations are not intended as a criticism of the Finnish language, or even the national tendency to make Finland even more Finnish. What I do criticize is the perpetuation of a bi-lingual myth which is in direct conflict with reality. Finland can no longer pat itself on its back for its liberal bi-lingual Finnish-Swedish attitude — when, from all points of perspective, this attitude has changed.

Outside Domination

Prior to the twentieth century, Finland was historically bounced like a soccer ball between domination by Sweden and Russia. First it was one side, and then it was the other side grabbing off bits and pieces of whatever was considered as economic or geographical prize from the land and its people. Oh yes, most dignified historians will proclaim in their museum voices that it was precisely this outside Swedish-Russian domination that eventually brought "civilization" to the otherwise romping aborigines, but this is only one side of the story. It must not be overlooked that the principal interest of outside domination is to *take from*, rather than to *give to*. It was therefore not until Finland gained its independence that Finnish self-interest could be securely protected from within. Consequently, already as late as the twentieth century — Finland discovered itself as a nation, and discovered itself with the problem of establishing an identity. It is no easy task for a nation to build a self-identity when the circumstances of its pre-history tend to pull apart all of its claims to cultural uniqueness.

A Twentieth-Century Phenomenon

What is Finland? Well, for the most part, and aside from its folk traditions, Finland seems to be a twentieth-century phenomenon. Out of a pre-history pocked with the scars of endless rape, Finland found itself a nation — it found itself as a nation after the industrial revolution and after it was already equipped with the tools of modern technology. Unlike its Scandinavian neighbors, Finland did not evolve bit by bit backed with several centuries of cultural tradition. One day back in 1917, Finland just hung its sign out, and began doing business as a nation. On that same day, it not only found itself with a need for an identity, but with hundreds of other cultural needs including an architectural style, and an object image. When evaluating the Finland of the 1970's, one tends to forget Finland's abrupt and recent birth. As a nation, Finland began from point zero so recently that there are still several hundred thousand living Finns who helped to celebrate her first birthday. What I believe all this leads up to is yet another reality of Finland — a newness — and with the newness, a healthy capacity to realize change.

Newness

There are of course disadvantages to newness. In itself newness tends to eliminate alternatives. When everything is more or less new — then there is almost nothing old. If objects are all mostly new — then there is almost nothing old. If traditions are all mostly new — then there are few that are old. Finland is therefore both blessed and burdened by its newness. It has the fresh capacities of youth, but yet it notices the absence of age. Compared with its Scandinavian neighbors, Finland has a minimum in handwork tradition, and a maximum in industrial tradition —it has a minimum in architectural tradition, and a maximum in the architectural by-product of modern materials and technology. Assuming that alternatives are healthy for a nation, Finland must therefore take care to deliberately *build-in* the alternatives which its neighbors have accumulated as a natural part of their ageing process. I believe the absence of alternatives, which results from newness, is another of the realities of Finland.

People

And what about the Finn? Are there really any generalizations one can build about the national personality of the Finn? If one listens to the stereotyped comments about the Finn — then one immediately assumes that he is not far removed from the forest. He is depicted as a character who drenches himself in alcohol and is too quick with a knife. He is painted as a brooding fellow — a character who is tight lipped, locked up internally, and a kind of primitive who doesn't mix too well with foreigners.

I have also heard outside comments that when you meet a Finn — you do not meet the inner man — you meet a theory. I assume that this is advanced as another indication that the Finn is closed — he doesn't let you see what he is inside, he simply masks himself in ideologies to the point where you know what theories he has, but you do not know what his joy, pain, and suffering amounts to inside. These generalizations about the Finnish mentality are particularly noticeable elsewhere in Scandinavia, and most of them are even well-known at home.

Rather than comment directly upon these stereotypes, let me here draw some analogies. Is it fair to generalize about the American and stereotype him as a fat-bellied tourist with Bermuda shorts supporting a thick billfold of dollars and a Kodak Instamatic camera? Is it fair to generalize about the Italian that there is something intrinsically evil behind his dark eyes, and his pointed shoes? Is it fair to stereotype the German as a creature who wears only high-healed leather boots, who clicks his heals together, and who believes in his own superiority? Obviously, when you begin to set these national stereotypes down on paper, they look ridiculous. Sufficient to say that for every single Finn who lives up to his stereotype — there also exist ten thousand exceptions.

It is perfectly ridiculous that the Finn should be characterized as a primitive, when as a matter of reality, he is a personality in one of the most modern, industrialized nations in the world. Yes, there are closed and brooding Finns, but so also there are closed and brooding Swedes, Germans, Americans, Africans, and Chinese. Yes, there are knife-wielding Finns, but so are there gun-toting Americans. One cannot generalize about a national mentality by drawing attention to its exceptions.

If there is an insight to be gained by making broad generalizations about the Finnish personality, about as far as one can honestly go in this respect is to suggest that the Finn has a tendency to be more non-collective than collective. If it is true that a Finn values his independence, then it is also true that when one is independently spirited, he quite naturally will have communication difficulties with others — even of his own kind. This may be what outsiders have labelled "closed", but it results from the preference for non-collectivism over collectivism.

There is no question that a Finn is particularly, and I believe, enormously hospitable to foreigners. He welcomes outsiders with the natural human curiosity of one whose land mass places him outside the mainstream of urban Europe. This generosity toward foreigners does not sound like the personality of the locked stereotype who is depicted as uncomfortable among strangers.

Perhaps it is true, as many people have said, that unless you are a Finn, it is difficult to understand all things Finnish — but herein lies an important key which suggests that even despite its circumstances of origin, there well may be many unique characteristics about Finland which keep it from fitting within the standardized mold of twentieth-century "western" progress. Many observers would criticize Finland because it does not exactly fit the mold of Scandinavia, or even the rest of Europe, but let me here suggest that these characteristics might well be precisely those which should be most preserved and protected.

It is currently fashionable within intellectual circles to promote the universal oneness of mankind — or a kind of new third world as Herbert Marcuse suggests. What this philosophical idealism means is not exactly clear to anyone — since there has thus far been no historical perspective gained with which to evaluate what has not yet happened. If this means that all of mankind should become like-minded — a kind of smiling and agreeable xerox copy of all others — then idealism is showing its dirty underwear at the President's ball. If like-mindedness is to become a symbol of human virtue — then isn't it also possible that mankind will literally grow bored to death with itself? I'm not at all opposed to this idealism of universal oneness. I believe it has its place. It obviously has its place within the context of America during the 1970's — where there seems to be first of all a desperate need for at least a regional togetherness among liberal thinkers. I do not, however, feel that this idealism applies in every situation — and I can frankly see no reason for its application in Finland. Perhaps this makes me sound like a tired political conservative — but so be it. Universal oneness is a fine piece of philosophical idealism, but does that mean it has to be imported to Finland? If universal oneness is anything like the universal Americanism which has spread itself even to Utsjoki — then I'm afraid of it. Just because elephants are useful in India — does this therefore mean they should be imported to the streets of Helsinki? For what reason should a Finn therefore be like everyone else?

Summary

What is Finland? We have touched upon a few of the realities which constitute Finland and the Finnlander. Many of these are not found in the tourist guides. We have suggested that Finland experiences a time gap from the rest of urban Europe because the Baltic Sea cuts her off geographically — we have suggested that she is a country of climatic extremes resulting in a kind of schizoid role for her people; we have suggested that because of her pre-history, she has experienced difficulty in establishing her identity; we have suggested that she has a language which isolates her from her neighbors; we have suggested that she is mostly a twentieth-century phenomenon conscious of its youth, noticing its absence of age, but yet highly adaptable to change, and we have suggested that the Finnlander has a tendency to be more non-collective than collective, but still highly receptive to outsiders.

As a result of this very quick, and admittedly incomplete survey, we have suggested that there may well be many characteristics of Finland and the Finnlander which do not in fact fit the standardized mold of so-called "western" man, and we have suggested that these are the very characteristics which, during a time when the popular cry is for like-mindedness, should be most preserved and protected. This survey has therefore provided a feeling for the realities, as I see them, which Finland must confront as its identity raw material. This is the ground environment out of which the objects of Finland are produced, and in which the object-maker must work.

It is not at all unusual to build a cultural, or identity base as the groundwork for evaluating the objects of a given nation, but in the case of previous efforts on Finland, particularly an effort called "Modern Finnish Design", I believe this groundwork was built with chronic nearsightedness which resulted in a noticeable lack of perspective. No matter how convincing an author may be, he simply cannot account for a nation of object producers only by placing them in perspective to birch trees and Akseli Gallen-Kallela.

On one of my first visits to Finland I was given a courtesy copy of an English-language Finnish magazine published by the Finnish Travel Association. In that magazine, I read a number of articles which were slanted to create the impression that Finland was a mini-copy of the United States, and that Helsinki was the London, Paris, and New York City of northeast Europe . . . oops — excuse me! Please bear with me a moment while I digress in order to relate something that just happened while I was writing this page.

Right while I was sitting here, the Postman delivered a multicolored picture postcard. It was from my mother and father and postmarked "California, U.S.A.".

My parents are visiting California on winter holiday. The front side of this postcard is divided into three equal sections illustrating three different scenes.

The top photograph illustrates the entrance sign and the driveway into the "El Camino Motel" located in Carlsbad, California. The entrance sign is yellow and red. It is an electric, neon sign, and it flashes on and off. In addition to the name of the motel, the sign reads: "Member of the Golf Course Inns of America". The photograph shows a new, grey and black Cadillac automobile parked in the driveway.

The middle photograph on the postcard illustrates the inside view of a "typical" El Camino Motel bedroom. This is designated as a double sleeping room — overnight accommodations for two people. The room appears to be large enough to house a normal family of four. There are two bouquets of plastic flowers in view. Both bouquets are red. One is standing on a desk and the other stands on a "night" table. The room has wall to wall green-flowered carpeting. The photograph shows a television set in the room. The set is turned on. There is one double-bed in the room. The headboards on the bed are shaped like two royal crowns and appear to be covered with synthetic leather, framed by gold-painted wood. The bed appears to be over two meters wide. There is a coin box over the bed. It appears to be one of the electric vibrator beds found in many American motel rooms. This is a bed with an electric vibrating motor built into the mattress. Before you climb into the bed, you are supposed to insert a 25 cent coin into the coin box on the wall. This coin starts the vibrating motor in the mattress, and you are electrically rocked to sleep.

The lower photograph on the postcard shows the putting green of the motel golf course. Two men are putting on the green. They are wearing golf caps. Two wives stand looking on. They are content. They smile. There is an electric, battery-powered golf cart parked at the edge of the golf green.

Without exaggerating, I have just described this picture postcard which now rests in front of me. My parents have mailed me similar postcards every winter for the past fifteen winters. The postcards have always illustrated motels where they have been sleeping while on holiday. These postcards have come from California, Connecticut, Florida, Louisiana, Alabama, New Mexico, Arkansas, and Minnesota. My parents have toured America via the American motel room. Motel America is a contemporary symbol of the affluent society — a society which has arrived at such a point of sophistication that it now has a choice between an electric vibrator in the mattress, or a pillow equipped with stereo speakers.

The shocking reality is that my parents, like millions of other "affluent" Americans, actually BELIEVE in motel America. Motel America is now an established part of the affluent value system. Motel America is the reality of where America, the super-nation, is really at.

Is this where Finland wants to go?

. . . oh yes, excuse me for my digression. Before I interrupted myself, I was telling you about an English-language Finnish magazine I read which was comparing Finland to the United States, and suggesting that Helsinki was the London, Paris, and New York City of northeastern Europe.

You will remember that I began this chapter by suggesting:

FINLAND IS FINLAND

and then I suggested two more slogans to hang up in the sauna:

FINLAND IS NOT THE UNITED STATES THANK GOD
HELSINKI IS NOT NEW YORK CITY THANK GOD

Now there is one enormous advantage for an American who writes a book about Finland: the author of such a book — me — although he may not know everything about what he is writing, certainly knows something about where he has come from. I have suggested that Finland *is* Finland because I see no reason for it to pursue an identity outside itself, and other than the one it has. But when one confronts the reality of contemporary Finland, and then reads a whole magazine which attempts to identify Finland for what it is not — then one becomes frightened at the prospect that perhaps it IS POSSIBLE that Finland really does desire to repeat the American experience.

For whatever it may not be, America at least stands as a historical monument to self-advertising — the advertising of its own mistakes.

America used to be the "new" country, and Americans used to look upon the countries of Europe as the "old" countries. But now the coin has flipped onto the other side. The American in Europe now sees the modernization of the "old" countries being formulated on the groundwork of the American experience. And he is frightened — and disheartened at the prospect. He sees countries which used to be themselves becoming countries not unlike his own. His natural instincts are to scream at the top of his lungs: "STOP"!

Yes, the American may well be a member of the richest nation in the modern world, but he also knows that the price his country has paid to gain its richness may well be its greatest mistake. The American (and I'm again speaking about myself), is completely confused when he sees the wheels of European progress aimed at duplicating his own mistakes. He cannot understand why it is that a nation like Finland, which has the option to follow its own dictates, should direct its energies toward mimicking his own culture. This becomes even more confusing when he realizes that through self-advertising, America has made its mistakes intimately known even in Helsinki. How can it be that much of modern Europe, and at times much of contemporary Finnish energy seems directed toward jumping blindly onto the bandwagon of the American experience — when in reality, that bandwagon has gone off its course.

So again, what is Finland? Finland is a land mass of energetic people lying on the OUTSKIRTS of Europe. Yes, it experiences a time gap, a language gap, and many other characteristics which at first glance might appear to be disadvantages during the fanaticism to become modern. But let me here suggest that Finland is in a unique position just because of what it is — by virtue of where it is. Finland has the unique opportunity to sit back and observe the insane rat race of "western" man without being dragged in by the force of the pull. Finland sits in a position where it can study the events going on around it, reject those which do not apply or do not bring rational benefits — while at the same time absorbing and instigating those which do apply, and which more or less guarantee rational benefits.

So if you believe Finland is helpless, and cannot protect itself from being dragged within and under the wheels of "western" progress? — then just ask yourself what it is that forms the whole groundwork for the Americanization of Europe. The answer is obvious — it is an ability to pay CASH. If a nation has the ability to pay CASH in the international game of politics and commerce — then it can keep its own identity — regardless of whether that identity fits into the rest of the machinery. All of the other trinkets that constitute the Americanization of Europe are really only decorative frills — like the lace on a petticoat. The Finn can do business as he damned well pleases — in his underwear, or in his fur hat — if he has the ability to pay CASH.

One final slogan to end this chapter:

"TO HELL WITH INFERIORITY COMPLEXES IN FINLAND"

Chapter 3
The Communications Gap

In the early years after the Second World War, Finnish industry was experiencing a recovery period — it was just then planting its economic feet solidly on firm ground. This was a re-building and expansion period during which the industrialist was desperate for new ideas. He looked around for a source of ideas and discovered a small group of professionals — Ateneum graduates for the most part — who then didn't even have a single professional name under which to group themselves. Some of them were called interior architects, others were called form makers *(muotoilijat)*, and still others were called functional artists *(käyttötaiteilijat)*. They represented youthful energy, and they had ideas. The industrialist recognized this, needed them, and aggressively pursued their cooperation. It was during this early recovery period that the two groups formed an intellectual working partnership and began producing objects which put industry back onto the healthy side of the profit columns.

For the most part, this was the "golden era" for Finnish object production, and during this period the people with the ideas began to be called "designers". In its initial stages, the working structure was so loosely organized that these idea people hadn't yet been grouped under the single title "designer".

Fate and circumstances were indeed kind to this first group of Finnish designers. The ideas they professed were put into reality, and Finnish industry transformed them into the first group of super design heros. Today, almost two decades later, these same people are now looked upon as the "father" figures of their profession. They got in when the getting in was at its most rewarding point, and most of them have stayed in — SOLIDLY. The early opportunities were endless, and it was not then uncommon for a young interior architect to find himself with opportunities in glass or ceramics. A young form maker *(muotoilija)* might also have found himself being asked to perform in wood, silver, wax, and even in graphics.

This first group of designers, because they were fortunate enough to be at the right place at the right time, established themselves in positions of leadership — positions which many young designers would today call a monopoly.

But the environment which existed for these early designers of the 1950's is not the environment which exists for the Finnish designers of the 1970's. Things have changed, and many of the changes have not been positive. The most noticeable change is a communications gap which exists between these same industrialists and the men and women they now employ in positions of design. Yes, the industries are now larger than they were in the early 1950's, and there is the inevitable increase in bureaucracy that always tends to frustrate attempts at communication, but this is not the major sore spot.

I believe what's happened is a reversal in energy source resulting from fat and healthy bank accounts. In the late 1940's and early 1950's, the industrialist *pursued* and *needed* the ideas of the young designer in order to secure his profits. The partnership worked, and was healthy. While the industrialist deposited his profits, the designer read about himself in magazines, and was being solicited to perform in still other areas. Perhaps the initial communication used to be something like this: THE INDUSTRIALIST ASKED. But today, after almost two decades and a fat bank account: THE INDUSTRIALIST TELLS.

There is an enormous difference between ASKING and TELLING. It indicates that there is either an oversupply of designer talent, a limited space for exploring new ideas, or a general laziness or confusion existing in some part of the production mechanism.

At this point I must align myself firmly on the side of the young designer of the 1970's, and leave no doubt about it. I feel that the young Finnish designer of today has talent and energy equal to his predecessors. In many cases he is bursting with talent, and as a matter of fact can often outdistance the skills of the "father" figures within his profession. The problem is that his talents and energy are being cut off. He is only realizing a small percentage of his capacity. He simply does not today have the same opportunity that was experienced by the designers of the early 1950's. When industry shoves a young designer into a pre-established slot and then TELLS him precisely what it is he is supposed to do — without allowing him freedom to fulfill himself — he is being cast in a role where he performs like one of several thousand other gears in an enormous, ambiguous machine. In most cases, this did not happen to the young designers of the 1950's, but it is happening with the designers of the 1970's.

One quite often hears comments today, even from educators, that young designers are instinctively lazy — as if the white corpuscles of the 1970's were suddenly eating up the red. I believe this is a lot of nonsense — these young minds have a surplus of energy. I think it is more truthful and closer to the point to say that many of the industries of the 1970's have grown *idea* lazy. A great many industries realized their success during the 1950's, and too many of them have been riding this success for over a decade without reinvesting energy. The forward thrust of industrial energy during the 1970's is not at all the same as it was during the 1940's and 1950's. There are countless examples. The typical example is the firm that began producing a line during the "golden era" of the early 1950's, and is still producing exactly that same line in the beginning of the 1970's. These firms have managed to stretch success to its maximum breaking point without heavily reinvesting. They still manage to ride on the Finnish reputation of excellence, and have made very few recent contributions toward insuring this reputation.

In the meantime, and while the industrialist has been counting his profits, the world outside Finland has changed. Other, more competitive nations have come to the forefront. With them, they have brought along fresh new ideas, exciting color, and the same youthful energy that Finland invested in during the 1950's. These competitive nations have introduced bold new forms — while back home in Finland, many of the forms have changed very little. What does this indicate? It certainly seems to indicate that the old reputation of excellence has been pushed to its ultimate limits, and that it may soon be in danger of noticeable cracks if there is not a general youthful rebuilding of tired blood.

Let me here cite several specific examples. Several years ago a young Finnish girl came up with a double-bench form for children. In my estimation, this is one of the most exciting open-ended children's forms to come along in many years. It provides the child and the parent with unlimited alternative uses, and it is a form which allows a child to use his own imagination. The form could have been fabricated in plastic and retailed inexpensively for a broad, general audience — this was the designer's original idea. For some reason the form ended up in the hands of the firm that has been producing Alvar Aalto's forms since the late 1930's. This firm works exclusively in wood, and therefore this double-bench form was fabricated in wood. It came out in Birch plywood — eleven layers of wood laminated together and bent to form under pressure. The fabricating process was difficult and costly — several times what it would have cost if the same form had been realized in plastic. The firm produced only a bare minimum of these forms in wood, but even with this minimum, the form was exceptionally well received both in Finland and in the United States. One would have thought that this positive reception would have encouraged the firm to further explore the form and its fabrication — but this did not happen. As a matter of fact, just the opposite happened — interest in the form from within the firm began to die. By the time I began photographing material for this book, there was only one example of the form remaining in the Artek showroom. This example was so badly abused that it was unsaleable. I made inquiries into the reason for the neglect of this form and was told that the firm management was much more interested in keeping the Aalto forms of the 1930's alive than in exploring what might have been a strong contribution to the forms of the 1970's. This of course was a management decision — one they were free to make — but in squeezing continued profits out of the Aalto forms, the firm was also issuing a severe slap in the face to the youthful energy of the late 1960's.

This second example seems even more bizarre. During the later part of 1970, I visited a large firm producing textiles. I sat for many hours in the design department talking with the people who were responsible for the firm's textile imagery. I immediately noticed that all of these designers seemed to be exclusively drawing dainty flower patterns. I was of course surprised to see so many small flowers — many of these people were barely over 20 years old — and I could not understand this youthful preoccupation with flowers. I inquired about this point and was told that, as a matter of fact, all of them were bored to death with flowers. How was it then that they were drawing flowers? Oh, one of the sub-chiefs in the bureaucracy over them had simply given them instructions to draw flowers. He had told them that the firm didn't want any of this modern stuff — they just wanted flowers.

In direct contrast to this non-creative flower drawing for the firm, many of these same young people were drawing fantasy forms of all kinds at home — but this off-duty imagery was unseen by the firm. About two weeks after this experience, I visited with a management friend of mine, who by coincidence, happened to be a personal friend of the director for the firm involved. I related the events of my visit, and my friend found my story incredible. He told me that the director of the firm did not want flowers — and as a matter of fact was sick and tired of flowers, and that flowers were beginning to hurt the firm's sales experience. He wanted fantasy — the public wanted fantasy — but all he was getting was flowers. He had been told by his management that his design department was dead, and he was seriously making plans to import fresh designer talent from Germany.

Here was a situation where the communications gap was at its greatest break. On the one extreme, a very alive design department was being frustrated by management in its attempts to break away from flowers. On the other extreme, a director sat with consumer support demanding fantasy and an end to flowers. The management bureaucracy in the middle was rapidly destroying both ends. Conservatism reigned as king in a situation where fantasy represented the pulse.

Still another example involves a young man who had come up with a very rational and very functional new form, a child's first food bowl. The form was designed with sides, and had a pronounced interior lip that extended back inside the bowl.

As the child brought his spoon up along the side of the bowl, the projecting lip kept the food from falling out of the spoon. The idea ot the projecting lip may have looked like a small point, but yet it represented a noticeable improvement over existing forms. The simple addition of this interior lip aided the child in its first attempts to feed itself.

When the initial idea was first submitted to his superiors, the form was ridiculed. It was called a dog dish, and the young man was obviously humiliated. He kept on believing in his form, but by the time it was finally accepted by the sales and production people, it had lost most of its original meaning. It was finally produced in a breakable ceramic material (the first mistake), and by the time it reached the production stage, the functional lip had all but disappeared. All along the route — from drawing table to production — there were bits and pieces of the original idea sliced away. When the form finally reached the markets, it was just another bowl, but production had managed to spice it up with colorful painted animals. In this case, the animals proved a poor substitute for the loss of a fine form. It is indeed sad to realize that in many Finnish industries, the sales and promotion departments have gained the power to dictate form to the man on the street and the man who is trained to know better.

A curious reversal in roles has occurred with this 20th century phenomenon known as the industrial salesman. By no stretch of the imagination is he the salesman he used to be.

In the old days, salesmanship used to be a profession requiring particular, special insights on the part of salesmen. When a man produced more of something than he could himself consume — he hired a salesman to sell it. This salesman was skilled at handling the public and he knew how to sell. You could give him a new automobile, a used fur coat, or a broken wrist watch and he could sell it. It was his business to study the psychology of both the public and the concept of buying. In those early days the cycle was completed when management produced a form, gave it to the salesman, and the salesman sold it.

But then came the sophistication of mankind and the birth of the industrial salesman. The industrial salesman became a new, specialized occupation. It was no longer the industrial salesman's concern to feel the pulse of the public — no, his job was to sell to another middleman, a private store owner, or a buyer for a large chain of distributing houses. Although this new phenomenon called the industrial salesman lost his intimate contact with the public — management has never really gotten around to realizing this, and in fact still today believes that the industrial salesman is the public link.

This has therefore resulted in the reversal of roles. The salesman no longer sells what management gives him to sell — oh no, management has elevated the salesman to a powerful position within the system. Management has placed the salesman on committees and on other boards with decision-making power over subjects which the salesman very often knows nothing about. The salesman now has a powerful voice not only in deciding what the management should produce, but in what form it should take, what color it should be, what texture it should have, how it should function, how it should be packaged, and how it should be priced. More often than not, the designers in these firms are not even consulted on these same points, even though the decisions directly involve their professional skills. Unfortunately, there are very few salesman who take the trouble to learn about color, about function, or about form. It therefore seems absurd that the salesman should participate in deciding upon these points in an often more powerful position than the men and women whose concern it is to struggle over these points every day of their working lives.

For the most part, it is the salesman's role to take the path of least resistance. On the industrial level he no longer has contact with the public, but yet his opinion is valued as if he did. He comes in from "the field" after having made contact with smaller businessmen — mostly businessmen who operate their own shops. These small businessmen are usually careful with their buying and their money. They prefer buying what is SAFE, and more often than not, they are out of step with the times unless they are fortunate enough to have contact with an aggressive public. This is the man the salesman deals with — generally, one of the most conservative elements in the buying-selling cycle, and certainly a great deal more conservative than the modern public. It is based upon this contact that management consults the salesman for advice on what to produce, and then gives him power of making decisions. Indeed, a curious reversal in roles. In the meantime, while all this curious reversal is going on — the highly skilled designer sits with an ever-widening communications gap. He is being transformed from a man with great potential for ideas — into a man who sits at a work table drawing flowers.

There is no question that the communications cycle is messed up, but there are many steps that can be constructively taken to repair the gap. One place to begin is for the designer to humble himself a bit, climb down from his aesthetic perch, and aggressively communicate with the salesman. He can also do a great deal more about keeping himself in contact with management. On the other hand, management can begin to pull back on the authority given the salesman, and recognize that he is no longer the public link he used to be. Management can also use the designer to help the salesman gain learning experiences on form, function, packaging, and other designer problems. There is a triangle that needs to be reconstructed from the inside-out.

The obvious benefit from this reconstruction is the likelihood that the public will receive better objects.

Additional specific examples of the communications gap are literally endless. After having interviewed several hundred designers, planners, architects, and yes, even salesmen and industrialists — the presence of the gap and the threat of it growing wider is indeed obvious.

It is not an easy task to develop new forms and to see them realized within their original integrity on the marketplace. In the early 1950's the industrialist was optimistic about taking a gamble. In the early 1970's, he has grown hard and much less explorative.

One seasoned interior architect spent a great deal of energy developing what he hoped would prove to be a furniture form answering the demands of young Finnish couples. He began the idea of developing forms that were simple to fabricate, light in weight, bold in color and form, forms that could be added onto, forms that were durable, easy to maintain, and above all — cheap. The form he came up with used laminated blocks of foam rubber that were sawed into sculptural shapes and then covered with a removable, zippered sack. He was so careful to protect the budget aspect of his idea that he worked out all the fabrication cost problems to the advantage of the producer. He even solved tremendous producer warehousing headaches by planning a form that was so quick and simple to make that it could be fabricated on customer order within several hours. Yes, the form was produced, but it did not come out according to the designer's plans, even though fabrication was carried out as planned. What happened was that somewhere during the process the producer made the decision to market this form as "exclusive" design. He realized that he had a potential winner — the form in fact WAS cheap to produce and simple to fabricate. He could have priced it for the young budgets, but he transformed the marketing into a sophisticated, design-conscious audience — selling the form at 2 1/2 times the pre-calculated normal mark-up. Yes, the producer was able to market the form, but in the process, he ignored the budget audience and squashed the designer's attempt at rational form with an economic conscience.

This growing frequency of a communications gap between designer and producer has also resulted in a noticeable gap between one designer and another. Competition is now severe, and has resulted in the growing tendency of designers to be even less collective. Individual secrecy and attitudes of competitive jealousy are now the rule, rather than the exception. When the designer loses his capacity to collect and group his strength with other professionals of his own kind, he loses a collective strength to repair the gap. It also results in an exposure to even further abuse.

Just after the Second World War, there was a great deal of idealistic optimism expressed in the hope that the industrialist and the designer could learn and grow from each other — while at the same time making joint contributions toward improving upon our object environment. These ideals got off to a very positive start by the early 1950's, but then breaks began to appear. The breaks have now grown into gaps and at times there is even a noticeable foul odor oozing up from many of the human relationships. No single party is entirely at fault. These ideals of the early 1950's are not impossible to realize, and inasmuch as such a healthy beginning was once under way, it now seems even more important to redefine those ideals and repair the gaps.

Chapter 4
The Image Game

The name of the game for the Finnish designer is the *IMAGE GAME* — at least the role of the designer often appears as if it were a game.
Here are the rules to the game:

1. Make on object.
2. Call the object either a "work of art", or call it "design".
3. Do not call the object what it is. Give it an exotic title. For example, do not call a clay object "ceramic" — instead call it "Fairy From Under The Snow". Another good one is: "Terrestrial Sunset".
4. Sell yourself.
   A. Do not call yourself a handworker. Handwork sounds banal.
   B. Call yourself either an "artist", or call yourself a "designer".
   C. Collect affidavits, certificates, and letters of recommendation from important people praising your genius.
   D. Get your name in the newspaper—or better yet, get your name on the radio or on television. If you are extremely clever and can arrange for key contacts — then try for the ultimate — get your PHOTOGRAPH in a magazine.
   E. Keep everything you say about yourself mysterious and exotic.
   F. Wear funny clothes, grow a beard, or shave off your beard.
5. Contact an industry.
   A. Show them how exotic you are.
   B. Do not talk about ideas. Talk about "art". Talk about "DESIGN"!
   C. Sell yourself.
   D. Sell your object.
6. Sit back and wait . . .
   and wait . . .
   for what?
   for image? No.
   for your object? No.
   wait for your 2—3 percent royalty.
7. Be content.

With the exception of rule number 3, which does not apply in the case of making furniture, these seven rules of the Image Game are not uncommon in the fields of glass, ceramics, jewelry, textiles, toys, decorative objects for the home, and many of the rules are even practiced by artists, sculptors, and architects.

Just in case you may have missed it, I should like to point out that rule number 1 requires the making of an object, but it is not until rule number 5—D (all the way at the end of the list) that the object again appears. The intervening rules are all concerned with image — the image of the person who made the object. If one intends to play the game according to the rules — then he must concentrate on himself first — the object is sold only after he has sold himself first. The object is not particularly important. The image is.

The frightening thing is to see this procedure for image success outlined on paper as if it were a game, and then to realize that if you play the game it REALLY WORKS. I have met dozens of Finns who have played quite close to these rules and have proved that the Image Game is valid.

In defense of the Finnish designer — the man who conceives the objects of Finland — I do not believe he is responsible for inventing this distasteful game. Yes, much too often he is required to play the game, but in all honesty, I believe he is an innocent victim. Who invented the game?

I believe the Image Game is an invention of the promotional and marketing people within Finnish industry and within governmental promotion and trade agencies. Without realizing it, or perhaps intending any harm, they have invented the game and cast the Finnish designer in the role of the public whore in order to market his objects. The designer has been cast into a role similar to that of the American movie hero — his private and public life has often become more important than his work. As a public hero, his objects are no longer objects. They are works of "art". They are status symbols. They are called "design", and very often they even have small paper labels glued onto them with the word "design" PRINTED IN VERY BOLD LETTERS. As works of "art", they therefore sell at a higher profit — a profit which does not reach the pocket of the designer, but a profit which remains tightly in the hands of the industrialist as his prize for marketing the object so cleverly.

From all observation, the public seems quite happy with having designer heros — even though it is the public, in the end, who must pay the cost concealed in the purchase price of what it buys. It is therefore not at all uncommon in Finnish homes to have a guest pick up an object and ask: "Who is this?" — rather than "What is this?". It is also not uncommon for a Finnish woman to receive a compliment about a piece of jewelry she is wearing around her neck by someone saying to her: "Oh, I see you are wearing a Mickey Mouse" — rather than telling her "What a handsome necklace". The Image Game has worked to such extremes of promoting the individual over the object that on a recent trip to Finland, and at a cocktail party,

I was obliged to literally swallow the name of the designer because it was prominently glued onto a paper label at the rim of the glass. It was very important for that particular individual to let everyone at his cocktail party see that he had glasses designed by somebody famous. It was so important that I actually felt intimidated by having to put the paper label in my mouth just to get at the contents of the glass. Perhaps the most common extension of the Image Game is to see the designer's private and social life reported in the gossip columns of the daily papers and gossip magazines — or to find that image building can be carried to such extremes that even a designer's baby is named "Finnish Baby Of The Year" — as recently happened (no offense directed toward the baby).

The Image Game works in Finland, but what the marketing and promotional people do not seem to realize is that when the same game is used outside of Finland, it makes the GAMESMANSHIP obvious — and in the end makes the Finnish image look a bit ridiculous. For example, in 1970 the shops on the Copenhagen walking street cooperated in a "Find Finland" week by placing promotional material supplied to them from Finland in their shop windows. Much of the promotional material was the same as at home — large photographs of famous Finnish designers. Some were smoking pipes, while others were just looking elegant or intellectual. These photographs had absolutely no meaning for Danes. The Danes were of course interested to see objects from Finland — but in far too many situations, they were being given only triple life-sized photographs of faces which had meaning only in Finland. Needless to say, from all Danish responses which reached me (and contrary to what was reported in much of the Finnish press), this promotional strategy was a complete flop.

The worst abusers of the game seem to be the huge monopolies such as Arabia in the ceramics field. Arabia employs a group of potters in a department which it calls an "art" department. Arabia calls this group of potters "artists". Arabia also tells us that what these people make is "art". We now have an enormous ceramic monopoly telling us what is "art", and in order to prove their point, they have established a kind of pseudo art gallery in which to exhibit this "art" by these "artists". I have looked in this gallery dozens upon dozens of times over the past 3 years. I see what looks to me to be ceramic handwork, but I am told that it is "art". I walk into a factory gallery and the factory tells me, and tells its public, that what is made in its "art" department is "art". The factory also indicates that these are "unique objects of art", but I am not fooled by this nonsense. I know that many of these objects are not unique — they are pieces which are part of limited editions — much like the limited reproductions of graphic artists. I also know that although some of these pieces claim to be made by this person, or that person — they are in reality made by assistants working in the same studio. I have picked up these pieces and have been shocked by the prices. Is this system of absurd pricing the industrial security blanket for proving that gallery objects are "art"? Whew, the whole thing is a nightmare and the sooner it stops, the better it will be for everyone. Factories defining "art", factories defining "artists", factories running art galleries, "unique art" which is not unique, prices which are ridiculous . . . all this fuss just to make a bit of handwork look like it was an act of God.

It may be of interest to Finnish readers to know that the Image Game is not the universal procedure in marketing objects. There are alternatives. There are dozens of nations involved in object production which do not use the personality of the designer as the marketing strategy. One obvious alternative is to promote the object. Switzerland, for example, is world famous for its watches. I've purchased Swiss watches, but I've never yet bought one that told me whether the designer smoked a pipe! In the final analysis, it is the object which really interests the consumer — not the personality of its maker.

One side-effect of the Image Game is that many smaller producers are now cashing in on the results of the larger firms who invented the game. These smaller firms do not have the budgets to promote the individual on a full scale, and so they take the next logical step by identifying themselves on the fringe of the Image Game with product labels that read: "Handmade In Finland", or even "Finnish Design". I know of one jewelry firm, for example, that stamps "Handmade In Finland" on forms that are cut out with a die machine and pounded with an electric air-hammer to make them look hand hammered. If an object made by a machine is now to be called "handmade" — then we had better change the dictionary.

The dangers in playing the Image Game ought to be obvious to everyone concerned. If the designer is to become a public hero — then he rises to a position which exempts him from healthy criticism. If he were working alone with his own hands this might still be all right as long as he maintained his personal integrity about the objects he made. But when he is a by-product of industry — there is no longer any way of insuring excellence. If it reaches an extreme, industry can use the designer's image to market garbage.

I suggest that the Finnish Image Game is out of control, and should be re-evaluated promptly. I believe the place to begin this re-evaluation is with the designer himself. If, in communion with others of his profession, he calls a halt, or at least a cooling off period in order to re-evaluate — then industry must follow his lead. In the end, industry can only get away with what the designer fails to object to. The solution sounds simple, but yet it is painful in practice. I see no reason for the designer to continue in his role as the industrial whore. Yes, it is true that industry gives the designer an enormous slice of ego, but ego unfortunately will not pay the rent.

Chapter 5

The Industrial Grip

At the time when the initial working relationship between industry and the designer was first formulated, both participants recognized and respected the individuality of the other. At that time, there was a more or less mutual understanding that to blend creativity with industry was not the same as to blend a normal factory worker with industry. In the beginning, the industrialist protected the individuality of his designer — he recognized that to create form was not the same human energy as to operate an electrical machine. From the outset, many industries therefore began loosely organized design departments where the designers were more or less free to come and go, and where they were provided with working environments and working conditions that were conducive to their special task of form giving. These early days were the days of implied trust and mutual respect. But then business got rolling, everyone got busy being busy, pressures began to grow, and then conditions began to change. The ideals of the early 1950's were lost within the hustle and expansion of the 1960's.

By the beginning of the 1970's, the Finnish designer found himself caught in the industrial grip. On the outside he was the public hero — but on the inside he was the industrial whore. He now realizes that his special position has been downgraded. In a few instances, he is even looked upon as a kind of misfit — as if his energies and justification for existence were a kind of necessary, but unprofitable evil. Although the designer still maintains his title, he has begun to lose control of his position. Whereas he formerly was an active participant in determining overall industrial direction, he now is only a tiny cog in a massive machine. His individuality has become ambiguous and his identity has become insecure. There is an enormous difference between being looked down upon as a necessary, but unprofitable evil, and being respected as a unique working partner. This is not to suggest that all factory situations are the same — thankfully, there are still those where the ideals of the early 1950's are still evident — but on the other extreme, there are dozens upon dozens of situations where the neglect and downgrading of the designer's position is only too self-evident.

To make the rounds interviewing designers about their working environment can at times become like an Alfred Hitchcock film. After hearing the first two or three dozen situations where neglect and industrial indifference have reached their maximum — one is tempted to crawl up to the top of Stockmann's flag pole and begin shouting names, addresses, and all the disgusting particulars.

The most common abuse is that of imposing an almost "Divine" loyalty to the firm. The designer is expected to be 1000 percent loyal to his employer, but there is nothing in the demand which requires the same loyalty from the firm. It is a one-sided bargain that is heavy to carry. To regulate its side of the demand, the industry establishes rules of procedure. These rules are imposed upon the designer. If he breaks the rules, it is clearly understood that he will lose his job.

Rule number one, and a rule which is almost universal in Finnish industries, is that the firm has the exclusive rights of ownership over every single idea, form, or nuance expressed by its designers. In other words, rule number one makes it quite clear that when you work for the firm — the firm owns every little hair on your chinny-chin-chin. If you examine this rule from where the industrialist sits, it appears only reasonable. The industrialist pays the salary of its designer-employees, and it therefore has a perfect right to demand exclusivity. Even from the designer's point of view, when the rule is expressed in theory, it doesn't sound too difficult to live with. But let us take a look at the way the rule is carried out in practice.

My first example involves an interior architect employed in a large furniture industry to design furniture. After several months of struggle, he had come up with a form. He showed me drawings of this form. I asked him what had happened. He told me that he had taken this form to his superior, but it had been rejected. He not only had drawings of the form, but had also made a life-sized prototype. The form was inexpensive to produce, it was quite easy to fabricate, it was highly functional, it involved open-ended planning, and it had been thoroughly tested prior to its submission. I asked the architect what had happened to this form. He told me that it was dead. I asked him if he had tried to sell it somewhere else. He told me that if he had tried to sell it somewhere else, he would have lost his job.

This same interior architect showed me drawings and a prototype for a lamp. His interest in interiors had followed its normal course into problems of lighting. He told me that he had submitted this lamp to his employer, but like the furniture form, it had been rejected. The lamp made use of the identical materials being used by the firm to produce its furniture forms, and it was in fact, a complementary form to one of the furniture series already in production. Just like the furniture form, it was a highly responsible form. I asked him if he had tried to sell it somewhere else. He replied that even though the firm did not normally manufacture lamps, and could not, under any circumstances, consider this as a competitive form, he was denied the opportunity of selling it elsewhere. The form died.

This next example involves a jewelry designer employed at a well respected Finnish jewelry firm. His job was to design jewelry forms. After several months of hard work, he submitted a whole series of new forms complete with drawings and prototypes. All of these forms were rejected. He tried still a second and third collection, but these were also rejected. The forms made use of the exact same materials and technical facilities being used by the firm. They were conceived for industrial production — exactly according to the capabilities of the firm. I asked this designer if he was free to sell his forms somewhere else. He answered in the negative and told me that everything he did was owned exclusively by his employer. He explained that if the firm did not wish to produce his forms — the firm therefore maintained a "Divine" right to kill them. A sad postscript to this example was that many months later, after the designer had left the firm, the firm began to produce the forms under another designer's name. I saw both the original drawings and the forms of several years later. There was no question but that it was plagiarism. Similar examples from jewelry designers have been repeated to me at least a dozen times. The jewelry industries of Finland certainly need to take a long, hard look at themselves as they misrelate to their designers.

I spoke with several potters who were employed by industry especially to design objects for industrial production. A few of them were painting stereotyped designs on cups — others were mostly making paper sketches and prototypes of dinnerware sets. I was curious to discover what these people were personally interested in by way of ceramic form. Several of them showed me sculptural ceramic pieces left over from their school days at the Ateneum. I asked if they were still exploring form at home in their spare time. Without exception, all of those I spoke with replied in the negative. I asked them if any of them had potting wheels or kilns at home. None of them did. I was told that industry did not allow them to make and sell sculptural pieces at home, even though their task at the firm was in the industrial design department, and quite unrelated to ceramic sculpture. I asked them what would happen if they decided to work at home. They all seemed to think they would lose their jobs.

I could recite several more pages of examples, but perhaps I have already made my point clear. Rule number one on paper is not rule number one in practice. It spells death to ideas and what is more, it stops the flow of ideas. This hard, and apparently unbreakable, rule of exclusivity practiced by dozens upon dozens of Finnish industries is diametrically opposed to the whole creative process. If I, for example, were to submit 10 designs to my jewelry firm and have all of them rejected — and then repeat this experience several times — then for what reason should I concern myself any longer with responsible form? What kind of an incentive is it for me to pursue responsible form only to have my idealism rejected? This number one rule of exclusivity is in practice the number one killer of creativity.

It strikes young people the hardest. It confronts them with the often cruel facts of reality when their creative energy is at its highest point. It forces them to come to terms with rule number two. Rule number two is to play the game. And what is the game? The game is to resign oneself to mediocrity when there is a potential for excellence. The game is to forget about classroom idealism of form with a conscience, form with open-ended function, inexpensive form, and form which provides solutions to problems of environment. The game is to grow up, and to face the cold facts of life. The game is to fit in rather than to remain an individual. The game is to discover what it is the industry wants you to design, and then to resign yourself to design it — no matter if it is over-priced, non-functional, or just plain rubbish.

For the designer who is not in an economic position to quit his job, the man with a wife and children to support, or a new car to pay for — then the game of rule number two is to see nothing, hear nothing, and to do precisely what you are told to do. Rule number two requires that if you want to keep your economic situation intact, then you must begin to wear the logo of your employer on your pajamas.

This is not to imply that all Finnish industries are guilty of practicing the industrial grip. Thankfully, there are many exceptions, and many firms which welcome and support responsible new ideas. The designer who reads this book will know exactly where he fits into this overall picture. There is no question in anyone's mind, however, that there is a desperate need for improvement and a restatement of the ideals which brought industry and the designer together in the first place.

Kaj Franck once told me that he had encouraged young people to live up to their ideals about responsible form by rejecting compromise. He felt that if each designer were to live up to his own individual standards of responsible form — then industry would have no alternative but to produce designer forms of excellence. This is the right place to begin. I would only like to add a collective postscript to Mr. Franck's suggestion. A whole design department collectively supporting principles of responsible form makes a much stronger impact than every man for himself.

This industrial policy of exclusivity needs some immediate revision in favor of the designer. It needs loosening up and liberalizing with fresh air pumped into it in order to secure the overall excellence of Finnish object production in the future. Yes, from the industrialist's point of view there may well be nearsighted advantages for exclusivity, but in the overall picture of Finnish object production, this policy stands in direct conflict to the whole concept of encouraging creativity. Here are a few possibilities to explore for realizing improvement:

1. Exclusivity be limited to a first-option basis on ideas and forms developed at the employer's expense. The firm employing the designer retains rights of first option on all forms developed within the firm. If the firm acts upon its first option and rejects the form — then the designer is at liberty to market the form elsewhere. Rights of first option are not at all uncommon in other business activities of industry. They are also common, for example, in the field of publishing. The designer should be entitled to at least the normal business courtesies demanded from industry in other areas of its business.
2. The designer should maintain full marketing rights for all forms which he develops on his own private time and which are unrelated to the products manufactured by his firm.
3. Industry should pursue a policy of assisting the designer to find markets for those forms which it deems valid, but which do not fall within the interests of the employing firm. This could be done on a split royalty basis where the industry and the designer joinly share any profits made from marketing employee forms on the outside. In other words, industry should assist the designer in finding a home for his ideas — the industrial incentive for this assistance being a share in whatever royalties are gained.
4. Whenever exclusivity is applied, it should be limited to a performance date. In other words, after an idea or form is presented, there should be a cut-off date wherein exclusivity is no longer valid. This will force industry to make up its mind about specific forms rather than to set them aside indefinitely. Perhaps a time limit could be set at six months, and thereafter, if the firm has not made a decision, the full rights of ownership revert back to the designer.
5. Industry should establish a review board with decision-making power to decide upon all problematic situations arising from questions of exclusivity. Such a board should have equal membership of both management and designer-employees. This board would arbitrate all situations which were borderline and which could jeopardize either the rights of industry or the rights of the designer.
6. In situations where the designer terminates his employment with industry, exclusivity be limited to six months after the termination date, and thereupon all forms and drawings not acted upon be returned with full rights of marketing back to the designer.
7. All situations where plagiarism is suspected be referred to the review board for action.

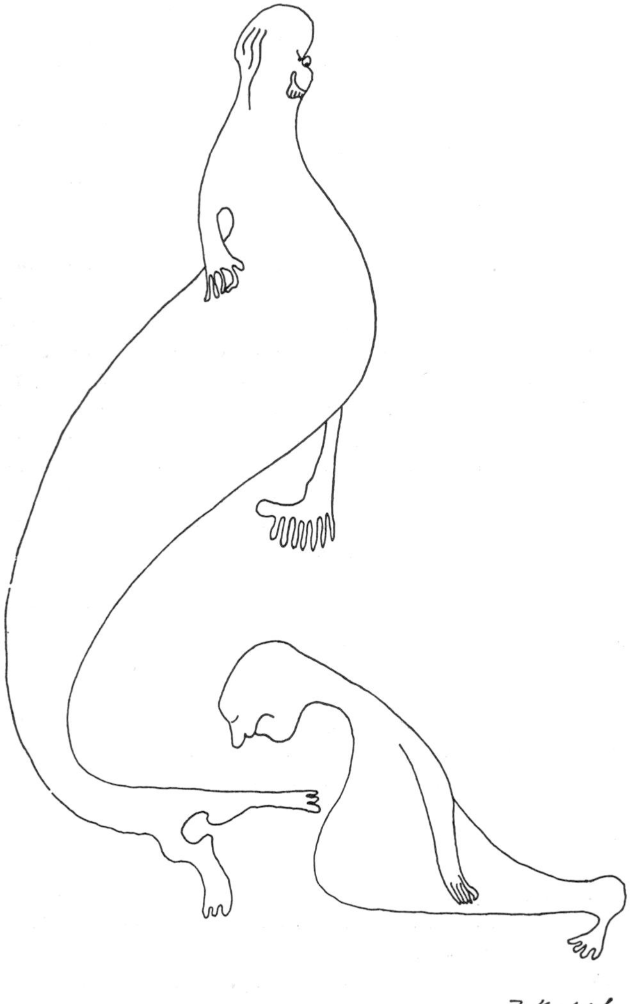

Chapter 6
The Economic Chasm

"We must only *pretend* to work here
when they only *pretend* to pay us".

The economic situation for the Finnish designer, and his inability to thus far protect his self-interests in an organized and reliable manner is indeed shocking to an outsider looking in. I would even go so far as to say that in many of the situations which have been personally reported to me by the designers involved, the economic abuses leveled against, and at the expense of, the Finnish designer are disgusting. The *Economic Chasm* which exists between the profits of industry and the earnings of the designer is perhaps the most fragile, yet explosive, point in this book. To expose the economic situation for the Finnish designer is to reveal him at his weakest point. The quicker this economic chasm is rectified, the quicker many other points discussed in the book will fall normally into place. The economic chasm is the first place to begin a major overhaul. Yes, my words are strong, but the situation is so chaotic that only strong words can be used. Rather than to spend many paragraphs relating my own opinions, let me begin by reciting a few examples. The only postscript necessary to underline these examples is that they are not exceptions. As hard as it may sound to believe, I did not go looking for these examples in order to hold them up as extremes. I do not believe they are extremes.

Here is an accurate reporting of the economic situation that existed for a young free-lance interior designer some $2^1/_2$ years after this designer's form for children had been on the market. The form had been well received by the Finnish press and had been photographically illustrated in the pages of every major design and home-oriented magazine in Finland. The form had recieved a design award from a children's toy counsel in the United States. After this enthusiastic reception, and $2^1/_2$ years after the form was available on the market:

1. The designer had received no contract despite many written requests.
2. The designer had been unable to commit the industry to an agreed royalty, despite many written requests.
3. The designer had received no money — not one penny.
4. The designer had received no statement of accounting as to how many forms had been produced and sold — despite several requests.

This second example involves an industrial designer, an Ateneum graduate, who was employed in the industrial design department of a large ceramics factory:

1. This designer, after intense and specialized training for four years, was employed at a salary of 200 Finnmarks per month to work 5 days a week on a normal full-time working day.
2. This 200 Finnmarks per month was before deducting taxes.
3. After remaining with this firm for two years, the monthly salary was raised to 500 Finnmarks. Also before taxes were deducted.
4. This example happened after 1966.

This third example involves a young free-lance jewelry designer who was asked by a Finnish jewelry firm to design a new series of forms for industrial production. The designer agreed, worked up detailed drawings, and even supplied one-to-one scale models of each form in silver. The material was submitted to the firm. After six months' waiting:

1. The designer was unable to get a decision from the firm.
2. The designer was unable to get the drawings or prototypes back. The firm advised that this material had been sent to Germany for study.
3. The designer was unable to get even reimbursement of expenses incurred for prototype silver.
4. The designer had received no money.
5. The designer suspected that the forms had been sold to a German firm and were being produced in Germany, but had no private funds to employ a lawyer.
6. The designer will not again work for industry.
7. This example happened after 1969.

A fourth example involves yet another young jewelry designer. This designer's forms are well-known all over Finland and are today available in most jewelry shops. The designer's forms are manufactured in series by a small industry. The designer is employed in the firm on a regular monthly work schedule.

1. The designer is paid 600 Finnmarks per month. This is before taxes are deducted.
2. For this 600 Finnmarks, the designer works 5 days per week in the firm.
3. Despite the fact that several thousand of the designer's forms have been marketed, the designer recieves no royalty.
4. Although the designer recieves no royalty, the firm believes it has been completely fair because it has provided the designer with a public image.
5. This example happened after 1969.

This fifth example involves a free-lance interior designer who has been in the business of designing furniture forms for over 10 years. At the time of my interview with him, and after at least 12 of his forms had been successfully marketed both in Finland and for export:

1. The designer theoretically receives a royalty of 2 percent.
2. This designer had thus far never received an accounting statement from any of the firms producing his forms. He had no idea how many forms had been manufactured and no idea how many forms had been sold. He had many times requested an accounting statement.
3. After 10 years and 12 successful forms, he had received a gross royalty of 900 Finnmarks.
4. By the time he had made his prototypes at his own expense, perfected them, and found a producer for them — he had lost money.
5. Thus far no firm had offered to pay even for the cost of materials involved in making the prototypes.
6. The designer lost interest in designing responsible form for industry.

This final example involves a free-lance Danish wallpaper designer who is well known to this author. He lives in a lovely countryside home with all the comforts of a twentieth-century gentleman. His economic struggles are at a minimum. This wallpaper designer derives the majority of his income from Finnish wallpaper industries. Because he is a foreigner, the industries assume that he has special secrets about markets and styles. In actual fact, the designer lives on a Danish island quite isolated from the contemporary "scene". He spends many weeks each year in Finnish wallpaper factories. While he is working for these factories, he receives:

1. 400 German Marks per day.
2. All travel expenses by air, to and from Finland.
3. All hotel, food, and entertainment expenses while in Finland.
4. It is not unusual for this man to work a continuous 30-day period in Finnish industry and return home with 12,000 German Marks in his pocket. He pays no Finnish tax on this earning.
5. This example is happening today.

After reading these six examples, you may insist that I have invented them. You may insist that it is impossible for a situation to exist where a talented Ateneum graduate is paid 200 Finnmarks per month working five days a week in a ceramics design department — and in contrast, a Danish wallpaper designer receives 12,000 German Marks for the same period of work in Finland. I invented none of these examples and was as shocked to hear them as you are to read them.

There are of course exceptions to these examples. For the designers who were fortunate enough to participate in the industrial expansion of the early 1950's, the economic situation is much brighter. Many of these older designers have healthy and stable incomes. For the most part, it is the young and the free-lance designer who is subjected to the most economic abuse.

It is a curiosity to note that the whole international reputation of Finland has managed to profit from the work of its designers, but yet the designer is virtually unprotected. The business of Finnish object production involves a number of magazines, has resulted in a number of books, and it supports many exhibitions, showrooms, societies, federal agencies, trade centers, and professional organizations. All this activity over objects, but yet the man who is responsible for creating the objects of Finland is economically neglected. He has been paid in image, but he cannot eat his image.

Obviously, there is enormous room for improving the economic insecurity of the Finnish designer. Here are some positive suggestions with which to begin:

1. All objects suitable for industrial production should automatically be subject to written contract upon request of the designer.
2. A regular designer royalty schedule should be established in Finland. This royalty figure should be included in all contracts. The royalty figure should be fixed — either on the basis of suggested retail price, wholesale price, or actual fabrication costs. I do not believe the ambiguous 2—3 percent royalty now paid by many Finnish industries is high enough. It is not uncommon for a Finnish author to receive a royalty up to 5 times greater than a Finnish designer. If book publishers can still manage a profit margin with a 10 percent royalty payment (in some cases even higher than 10 percent), then certainly Finnish industry ought to be able to substantially upgrade its present royalty schedules. Royalty schedules should be fixed by contract for both foreign and domestic sales, and should include a provision for the sale of foreign manufacturing rights.
3. In the case of free-lance designers, all drawings and prototypes accepted by industry for production should be subject to cash payments — either as advances against royalties, or as outright purchase prices, exclusive of royalty. This is the only means of insuring immediate economic return for time and material costs incurred by the designer while perfecting the ideas. The free-lance designer should not have to wait months or even years for the form to reach the markets before he realizes a return.
4. All contracts should include a provision to guarantee the designer's involvement in the fabrication, pricing, packaging, and marketing of his forms. This is one way of encouraging object excellence, quality control, and marketing honesty. The designer must have the right to see that his forms are carried through according to his plans. If the fixed royalty schedule is high enough — for example equal to the royalty paid to Finnish authors — then the designer should be willing to involve himself in fabrication at his own expense. In other words, if his royalty is commensurate with other professionals using royalty schedules, then he ought to be willing to help industry with the fabrication problems without receiving extra salaried payments for his time.
5. Contracts should include a provision whereby any money which is received by the industry for awards, or for reproducing the forms in publications, be equally shared with the designer.
6. Contracts should include a provision whereby, at least semi-annually, the industry must make an accurate accounting to the designer on paper — showing the number of pieces produced, the number of pieces sold, and the amount of royalties due. Royalties should be paid in cash semi-annually — no later than 30 days after the semi-annual accounting dates.
7. Contracts should include a provision whereby the designer has the right to audit all production and sales figures involving his forms, and to also audit all royalty accounts.
8. Contracts should include a provision whereby the designer is entitled to at least two free samples of all forms produced, and he should also be entitled to purchase additional forms at fabrication cost.
9. Contracts should include a provision whereby at the designer's discretion, his name be included on the form, and on all promotional material regarding the form.
10. Contracts should include a provision whereby the designer maintains the right to assign his royalty to whoever he so chooses.

If these suggestions are adopted and then added to the suggestions of Chapter 5 dealing with exclusivity and the Industrial Grip — then the designer will have established himself with the same normal business rights afforded professionals in other fields.

There are of course many ways one can go about establishing these working guidelines. They are not at all naive ideals.

The most constructive way to begin is through designer collectivism — a strong professional design organization with the energy initiated by the designer himself. If for some reason, the designer is unable to collectively gain protection for his individual economic affairs, then I am entirely in favor of the State coming to his rescue to secure his rights. The State could also require that all industries in Finland maintain accurate royalty accounts with production and sales figures for each of their designers. If the pressure came from both ends — then there would be no future examples of designers going on for years without accurate figures on the activity resulting from their forms.

It seems to me that designers themselves could also set up an advisory council — a council that other designers could turn to in the event of problems. This council could serve to encourage the placement of valuable forms by giving the designer suggestions on firms, suggestions on how to professionally present and prepare his ideas, suggestions on industries outside Finland that might be interested in particular forms, and even help with such small things as simple business bookkeeping, correspondence, and advice on contracts.

Chapter 7
The Advertising Myth

"The most beautiful candles in the world.
They grow in Finland. The Juhava candles.
They bloom whenever you wish.
Design Timo and Pi Sarpaneva.
Not Santa Claus."

... quoted from "Designed In Finland", 1971, page 18.
No matter how naive my plea may sound in light of the above quotation, I should like to call upon the conscience of Finland and the conscience of Finnish industry to abandon their advertising myths and set about to interject a portion of honesty in claims of greatness. It is my feeling, based upon past and present performance, that far too many myths are originating from Finland for its own economic safety. If either the products are not substantially upgraded to meet the claims of the advertising, or the advertising toned down to equal the products — it is not at all impossible that Finland will run into sales losses on the international markets.

Several years ago the Volkswagen Corporation of West Germany gambled on honest advertising — if not even a tendency to undersell — and it worked. As the pages of this book will indicate, Finland's own Marimekko, with the help of Timo Liipasti, also made a substantial step into honest advertising — and it also worked. There is no reason why claims of greatness cannot be toned down to agree with the facts.

This fine line between honesty and myth could be debated within several philosophical volumes — with still no answer. It is therefore impossible to set down principles of honesty. Perhaps honest image building begins within the ability to confront one's own conscience. If all of us are required to be as great as we say we are — then we will exercise caution in our claims. If the products of industry had to be as excellent as the claims they make, then either the products would be greatly improved upon, or the claims would come closer to the truth. And if Finland had to be what it claims to be, it would tomorrow cut at least half of these claims.

It comes as no surprise to anyone that the rule of self-promotion is to exaggerate — to stretch the truth to the breaking point. Unfortunately, honesty is the exception. We have progressed so far in our profit-oriented society that most of us already understand that honesty has been re-defined to mean exaggeration. We all accept this insanity as if it were normal, and then we look upon a puritan as if he were a circus freak. Somewhere it has to slow down. Why not begin in Finland? Why can't honesty become a part of Finnish design? And why can't honest design become a part of Finland? If there is still part of an identity missing in Finland, then why not fill it with honesty? Can anyone imagine one honest nation? If an honest nation were to come along in the midst of twentieth-century chaos, it would be the greatest contribution to mankind since the beer-opener!

When one writes a book like this, one grows very weary of watching the typewriter keys issue criticism. One grows even more weary of seeing himself cast in the role of a naive idealist just because he is in favor of a few principles of common sense in a world that has long ago lost its conscience. It would be a much more delightful task to report positive energy, but when one suggests room for improvement, readers want examples — they demand logical PROOF — or otherwise they shout ABSTRACTION, and can never see themselves as anything but perfect. So . . . here are some examples. Here are some examples of stretching the truth to its breaking point. And examples similar to these are still happening in Finland every day.

In December of 1969, "Avotakka" magazine published a spectacular design issue. That single issue of "Avotakka" represents the exact opposite of what this book is all about. In the past, and for the most part, "Avotakka" has been a highly responsible publication offering a maximum of consumer information and product comparison, but then along came the December, 1969, issue like a colorful bomb upon the design environment. The magazine reached me through the mail without comment from one of the members of the editorial staff. I could not believe my eyes, and my eyes told me there was hardly a single responsible impression in the whole magazine.

Although the magazine did not come right out and admit it, the issue was not the result of the normal "Avotakka" staff. It was a special big business issue — an issue sold to the wood and paper industries of Finland so they could use it as a promotional instrument. Pure economic profit had instantly changed a hard-working magazine into a gimmick. Instead of having the regular "Avotakka" staff produce a responsible issue, the wood and paper industries employed one of Finland's "famous" designers, a man who had arrived at such a point of greatness he could do no wrong — and who consequently missed no opportunity to feature himself everywhere in the magazine prominently. Without going into descriptive detail about what has already happened as old news, I can report without reservation that just about every conceivable slant on Finland was stretched to its maximum truthful breaking point. That single issue of "Avotakka" stands as if it were a monument to irresponsible publishing. The sad postscript is that industry and the public in general seemed to love it. One "Avotakka" editor reported that the magazine was deluged with requests for more of the same. If that issue of "Avotakka" represents Finland, then I have never been in Finland.

I believe another example of the advertising myth, and one that most readers will remember, was the early spring 1970 exhibition at Helsinki's Taidehalli, entitled "Finland builds bigger". Although I stood many hours at this exhibition, I could never fully understand its purpose. It appeared to me to be in direct conflict with human reality — the monthly reality of structural shortages, and skyrocketing rents. What the exhibition appeared to be was a society of Finnish architects patting themselves on the back and congratulating each other for their professional "genius". This kind of an exercise would have been quite all right if confined to a closed clubroom, but it seemed in very bad taste when the public was especially invited to attend the ceremonies. Not only had the whole of downtown Helsinki business visually endorsed the orgy, but they had participated in enticing the public to attend. Yes, this was an exhibition of structures — cold cement, steel, and glass, but where was the conscience of the architect for the people who had to live in these rectangular "works of art"? And where was the architectural concern for the dozens upon dozens of architectural needs that still exist in Finland?

One redeeming feature of the exhibition was that not even its director believed in it — even though the inevitability of holding the exhibition was somehow beyond his control. While this exhibition was going on, you will remember that a vivid protest exhibition was being held at the same time in the student house at Aleksanterinkatu and Mannerheimintie. Strangely enough, and just like the "Avotakka" example, the public and much of the press seemed to feel that this exercise of architectural ego building was a whopping success. A friend of mine bought several folding chairs left over after the exhibition. They were cheap. He was happy.

Yes, there are other examples, unfortunately dozens of them. The worst offenders seem to be that group of people who put together written texts with photographs and then publish the results in the form of books, magazine articles, newspaper articles, tourist brochures, advertising pamphlets, films, and even radio and television scripts. This book graphically illustrates one vivid example, which might in itself seem quite harmless, but yet is still stretching truth to its maximum breaking point when it advertises that even the flowers in Finland smell better than anywhere else in the world. No single stone seems immune from being turned over to discover maximum exaggeration underneath. Yes, dozens of other countries are also involved in stretching the truth, but is this license for Finland to join the abuse?

If one believed all that was written about Finland (God

forbid) — then one would assume that Finland was a land of milk and honey void of problems — a land of long-boned blond nymphs frolicking in perpetual sunset — and a land of designers sitting in forests waiting for their magical inspirational communion with birch trees. Is this the land that annually watches thousands upon thousands of its most vital sons and daughters exit to Sweden? Perhaps if half of the money spent on image building were invested in solving problems — then some of these many thousand discouraged Finns would remain at home in order to contribute to Finland's social and economic growth.

This endless enlargement of the advertising myth could be brought under rational control if the author, the publisher, and the photographer could begin to exercise restraint. These three groups of professionals are in the end responsible for most of what has thus far gone unchecked.

For example, most of us living in the last half of the twentieth century can thank these three groups for what a few social scientists have called the "universal physical inferiority complex" where every man, woman, and child on this earth has been duped into believing he is a bit abnormal because of inadequate height, inadequate weight, inadequate hair color, inadequate smile, inadequate teeth, inadequate bust measurement, and inadequate muscle structure. From the cradle to the grave we are bombarded with mannequins of perfection against which we are told to compare ourselves. We of course do not measure up. We are inadequate. Could it be that for all these years it has been the mannequins who were abnormal — while all of us were quite all right?

I only point this example out to indicate that if the author, publisher, and the photographer were to consider their responsibility to their public ahead of their responsibility to their bank account, the breakthrough of improvement would be well under way, and could sweep even the economic power of industry along with it, whether industry agreed or not.

It all begins on the individual level. I began this Chapter with an admission of what many will call my naive idealism. Is a plea for common, ordinary responsibility now to be defined as idealism? Have we mutated so far afield in our intellectual sophistication that a simple plea for responsibility is to be looked upon as if it were the wild ravings of a lunatic?

Chapter 8

Ateneum

The Ateneum (Finnish Academy of Industrial Design) is charged with the vital responsibility of securing the excellence of Finnish object production. Theoretically, it is the very heartbeat of object production — the single educational tool where by a student raw material is trained to conceive the forms of the future. The Ateneum of today should be one of the major devices for projecting a measure of Finnish objects in the future.

Much has already been written about the Ateneum, and it has long been the subject of controversy. There is no question that as an educational institution, the Ateneum building as it now stands bears all the external and internal signs of a medieval ghetto. The building stands as a daily reminder to an otherwise modern Finland that it has severely neglected its professional visual education. The wonder is that it took so unbelievably long for the public conscience to feel its responsibility.

The internal administrative problems of the Ateneum are a subject unto themselves. I will not evaluate these here. Instead, I will try to confine the evaluation to the conceptual level of an overview on the problems of visual education in Finland.

As of this writing, plans are underway to build a new Ateneum. Already the graphics department, the photography department, and the industrial design departments have been relocated into improved environments. In the hope that public neglect of visual education will not again occur in Finland, let us briefly review the situation that formerly existed, and still today exists for several of the Ateneum departments.

Here is what we have done. We have taken the Finnish student from the last half of the twentieth century — the student who is filled with both energy and optimism — and we have installed him in an institution that looks, feels, and even smells like it had just been carried down from great-grandmother's attic all covered with cobwebs and dust. We have plucked the hope of future Finnish creativity from its home environment, and we have given it an educational environment that advertises its own internal decay. We have charged this young talent with the responsibility of future creativity, but we have provided it with visual neglect and a shocking absence of public funds. We have bought and paid for a reputation of object excellence, but we have so badly neglected the talent that is supposed to uphold this reputation that it feels like it is at the very bottom of all priority lists.

And what about the students within such an atmosphere? Well, as I see it, they have had one of three alternative choices. One alternative has been to see nothing, hear nothing, and know nothing about the atmosphere that has surrounded them. This alternative requires a blind acceptance with all energy being diligently directed toward study. There have been very few students who have elected this first alternative. A second alternative has been to protest, to carry signs, to close the school, and to try to actively alert the public conscience. The third alternative has been to drop out — go home, sleep, make love, or smoke hashish.

Unfortunately, none of these alternatives have been directed toward positive visual education. All three alternatives place non-educational problems on the shoulders of the student. The student has not asked for these alternatives. We have provided them.

All of us are keenly aware that in the past, far too many students have followed alternatives two and three, rather than to deplete their energies on education. And strangely enough, when the student has either dropped out, or raised his voice in protest, many of us have criticized the result and have wondered what is wrong with this younger generation. Is it really a question of what is wrong with the younger generation? Or, is it a question of what has been wrong with the public conscience of the adult generation? When we present our young student talent with no choice but to either drop out, or protest, how can we then blame them for doing what comes naturally?

The one positive result of all this previous non-educational expenditure of student energy, has been that the students themselves have awakened the public conscience. It is not a happy postscript to realize that in retrospect, our own sons and daughters had to teach us to recognize our own educational blindness. Hopefully, in light of recent controversy on the subject of visual education, the blind spot will not recur.

And what about future Ateneum plans? Well, let me here record the reactions of one of Finland's most respected visual educators — an educator who must daily confront the real problems created by the educational structure and the environment within the Ateneum. This educator believes that the new plans for expansion have fallen as the unfortunate victim of power politics. She feels that the future of visual education is being manipulated somewhat like a rubber play toy. She feels that the new plans are frightfully lacking in open-ended alternatives, and that current plans may well prove that the results will be obsolete before the new structure is even ready for occupancy.

This is of course only one opinion. There are obviously many other opinions in support of current plans. The controversy surrounds a fundamental question that must be answered honestly before plans are realized. Is it the business of politics to plan and project future visual education — or is it the business of the professional educator, the administrator, and the student?

Long-range, open-ended planning is a difficult concept to sell to a public conscience that is primarily concerned only with plugging temporary holes to stop up the noise. Unfortunately, the only sure way of making the conceptual theory of long range, open-ended planning understood by the politician and by the everyman on the street is to make it economically profitable for the public at large, including the politician, the businessman, and the student. If there are Finnmarks of profit to be gained by open-ended, long-range planning, then the power structure responsible for making decisions will quickly understand an intellectual concept which otherwise sounds too theoretical. This small piece of advice — which has incidentally been practiced successfully for the past ten years by the Danish Institute For Center Planning — may well be worth vigorous investigation as it relates to Ateneum plans for the future.

Still another possibility for the Ateneum of the future is to vigorously investigate the advantages of decentralization. Finland now finds itself with one major school. This is located in Helsinki, and therefore Finnish students are obliged to live in Helsinki in non-institutionally supplied housing. If schools are to be considered as institutions for students, then student problems should be taken into account in future planning. It is self-evident that student living costs in non-institutionally supplied housing in Helsinki are much greater than costs elsewhere around Finland.

Yes, I believe decentralization should be investigated, but I do not feel this should be approached in the same manner as was done at Dipoli. Dipoli is architecturally both interesting and aesthetic, but it stands as a segregated institution. Students need to be involved in contemporary life — they should not be isolated from it. I do not believe the answer for modern educational housing is to simply clear a patch of Finnish forest and segregate students into it.

On the contrary, decentralization should be investigated from the point of view of establishing 2 or 3 other Ateneums spread out in the other major cities of Finland. To decentralize by making small branches of the Ateneum is also not the wise solution. Decentralization should be explored on the basis of establishing complete schools in these smaller communities, rather than incomplete, partial branches. If such plans are realized, then the students involved in visual education can at last have the housing which is today non-existent.

Perhaps still another aspect of long-range Ateneum plans would be to expand curriculums to include whole new departments. When the demand for industrial designers was at its high point, several Ateneum areas were lumped together into an industrial design department. The school quickly began training students to answer the demands of industry. Perhaps a better solution would have been to train students to participate in directing the path of industry rather than simply filling an industrial hole.

The curriculum might, for example, include a department which deals only with the social responsibility of the designer. Some kind of a department is needed in order to give the student an overview of the maximum rather than the minimum application of design. If students are given an opportunity to actively feel their social role, and to actively participate in educational strategy, politics, government, and even administration, they would then have some groundwork experience with which to realize their social role.

Another expansion of the curriculum might include a department dealing with problems of modern jewelry design. As of this writing, there is no place in Finland where a young man or woman can receive training in this area. Yes, he can attend a goldsmith school, but goldsmith schools are primarily concerned with technical problems in gold and silver. These schools are not aggressively concerning themselves with concepts of form, color, texture, non-traditional materials, and other visual considerations. In contrast to this absence of a school offering jewelry design, Finland is overflowing with small jewelry industries. From where do these industries expect to get their talent in the future?

Perhaps the quickest way to survey the educational situation for the creative arts of Finland is to compare the facilities in this field with the facilities offered in other academic fields. After only about a five-minute comparison, it will become exceedingly self-evident that students in visual education have remained near to the bottom as far as the expenditure of public funds. If the forms of the future mean anything at all to the hundreds of Finnish industries that live off of them, then dramatic changes are needed.

Chapter 9

Agencies, Centers, and Societies

Finnish Design Center

The Finnish Design Center exists in order to provide exhibition and display space for the objects of Finland. In reality, the Design Center has been the monopoly of only the larger Finnish industries. Those industries which exhibit and participate in the Design Center must contribute economically in order to keep the Center running. To become a participating exhibitor requires a sizable annual expenditure, and it is because of the size of this expenditure that smaller firms find it impossible to economically participate.

The Center has a reasonably comfortable exhibition space, and in the past, has tried to divide this space into both a permanent exhibition of stable objects, and a changing exhibition of new objects. For some reason, perhaps because the Center does have an aesthetic veneer, it has been considered quite sophisticated — more sophisticated, for example, than the Finnhand exhibition organized for smaller firms.

The Center appears to have three main functions — exhibiting Finnish objects to the Finnish public, exhibiting new lines of Finnish industrial objects, and providing a place where foreign buyers can survey Finnish objects. All three of these functions are of course limited to participating members of the Center.

On the surface, the Center looks to be healthy and thriving, but behind the scenes, it appears to exist from year to year on a precarious economic basis. I had hoped to be able to report something in detail about the Center's organizational and financial structure, but despite repeated requests and promises, the Center has failed to respond. I believe the lack of response is a direct result of understaffing and overwork rather than a deliberate unwillingness to cooperate.

In support of my suggestion that the Center suffers from internal economic troubles, let me point out that in the past two years, two of the Center's principal leaders have left — the Director found more advantageous working conditions with one of the Center exhibitors, and the Exhibition Chairman found opportunities personally more attractive in Sweden. To find the Center caught without direction for many months would seem to indicate an abundance of internal difficulty. On several of my visits to the Center, the staff has from time to time even shown doubt about whether or not the Center would be continued.

I personally believe the Center is an extremely important institution, and I would hope that prompt steps will be taken to secure its permanence. I also feel that the original conceptual structure of the Center needs some serious re-organizing. Something within the original structure caused the Center to be limiting and closed rather than non-limiting and open. If an institution has a built-in limiting structure, then it has a tendency to become inactive and boring. On the other hand, if the structure is re-organized so that it becomes non-limiting and open, it will gain vitality, and expand precisely because of the unexpected.

If the large industries supporting the Center are in doubt about whether or not their economic contributions are profitable, and if some of them are seriously considering dropping their future participation, then I believe public funds should be used to insure continuation of the Center, but only after it has been re-organized into an open-ended, multiple use institution. In no matter which structural and economic form the Center is continued, there is no question that its function can become much more vital. I offer the following possibilities for consideration.

In order to bring the function of the Center closer in tune with the pulse of change, the Center could begin to regularly include in its schedule, a series of open, non-juried exhibitions — free to both the contributors and to the general public. Yes, it is true that for the past several years Ornamo members have been able to hold annual exhibitions at the Center, but one exhibition limited to Ornamo members is not at all the same as a regular series of exhibitions where all participants are welcome to exhibit whatever it is they have manufactured, or they have produced with their own hands. This kind of an open, non-juried exhibition will encourage students, young people, and others with creative ideas to expose and express themselves. It is entirely possible that through these free exhibitions, manufacturers, designers, and even the general public will discover active creative energy that would otherwise remain unexposed.

The operating name "Finnish Design Center" suggests that this institution represents, or at least should represent, the pulse of activity in the field of creating form, but as we have already seen, it is an exhibition room mostly restricted to supporting members only. If the Center is to come closer to what its name suggests, then there is no reason why it could not become a place for lectures, educational workshops, films, happenings, and whatever other activities that form makers, and form manufacturers are concerned with. The central location of the Center is excellent, and by extending itself into these new areas, the Center could become an active rather than passive vehicle for promoting an exchange of ideas.

Another extension of the Center could be to occasionally use it as a contact medium with the general public — a place to expose new forms, try out new ideas, test prototypes, and gain public and critical reaction in advance of production. If for example, a firm is considering a new object for production, it could use the Center as a place to sample reaction before heavily investing in production. The public could be invited to try out the object, to register comments, and to give constructive, or even negative suggestions. Perhaps if the public were directly involved with the forms it would like to have, producers would then become more conscious of needs rather than embellishments.

Still one more extension of the Center could be to provide an occasional center for involving the public in art, handcrafts, music, and even amateur theatre. If, for example, for one month of each year, the Center were used to encourage people to express themselves in creative areas — areas they could not otherwise experience in their city apartments or at their places of employment — then the Center would indeed be performing a valuable public service. If musical instruments, work tables, tools, materials, and other basic necessities were available for a general public to try out and to personally experience, then there is no doubt that it would be a true center of activity.

These have but a few possibilities. There are many others. The Center could even serve as a place to explore problems of communication between people, a place to expose sensory reactions to touch, sound, taste, and sight — or even a Center for exploring design answers to social problems. Again, the point is that if the Center is to survive as a vital part of the Helsinki community — rather than as a special club for members only — then it must come much closer to the meaning in its own name: "Finnish Design Center".

Finnhand Exhibition

In June of 1969, the Finnhand permanent exhibition opened its doors in the Messuhallit-minihall in Helsinki. Finnhand began as the small producers' alternative to the Finnish Design Center. Initial rates for permanent exhibition space were set at 225 Finnmarks for the first month and only 125 Finnmarks per month for additional months. In exchange for this minimum monthly rental, the exhibitor receives approximately 4 square meters of exhibition space, a booth designed to hold an exhibit, lights to illuminate the exhibit, an exhibition hall that is very often open 7 days each week (particularly during regular Messuhallit exhibitions), and a competent and courteous staff to manage the exhibit and answer public inquiries. Finnhand also provides its participating exhibitors with business assistance, marketing advice, export advice, and even help with correspondence and the translating of letters.

Admittedly, Finnhand is not as spacious or as elegant as the Finnish Design Center, but elegance is not its objective. As an alternative to the Finnish Design Center, it hopes to expose the products of smaller industries, and even handcraftsmen — to both the general public and the wholesale and export markets. At the time of my visits to Finnhand, they were averaging between 60—75 separate exhibitor booths — with objects ranging from outboard motor propellers — to fur coats — to even dozens of objects made at home by hand.

As late as 1 1/2 years after its opening, Finnhand had received very little press coverage, and as a matter of fact, the editor of Finland's largest home decorating magazine had not even heard about the exhibition until I reported its existence. Perhaps part of the reason for the absence of an enthused press is related to the absence of Finnhand cocktail parties, and other miscellaneous aesthetic embellishments. While the Finnish Design Center is exhibiting the products of 15—20 large firms with a budget for embellishment, Finnhand is trying to manage 4 times as many producers with a budget for economy.

Frankly, I find the existence of Finnhand an encouraging exhibition alternative. Because of its very low rates, it gives exhibition opportunities to almost everyone in Finland (if space were only large enough).

Apparently, Finnhand and Design Center look upon each other as competitors rather than as co-partners in the same venture. This is indeed unfortunate, because in the end the public, the foreign buyer, and the producer all lose out. It would be encouraging if a foreign buyer visiting Design Center could be referred to Finnhand, and Finnhand visitors referred back to Design Center. This cooperative exchange would bring about a much closer realization of the goals shared by both institutions.

Finnish Society Of Crafts And Design

The Finnish Society Of Crafts And Design, perhaps more than any other organization in Finland, has been responsible for building the international reputation of excellence now enjoyed by Finnish objects. For this achievement — which has been no small task — the Society deserves an enormous standing ovation from both the designers of Finland and the industries which have enjoyed the sales results of this reputation building.

For whatever its organizational goals might have been initially, the Society has now become a kind of central public relations bureau for the objects of Finland. The Society is supported by public funds, and a large share of this public funding is reserved exclusively for both domestic and foreign exhibitions of Finnish objects. When the major portion of funding is reserved exclusively for exhibitions, the Society is therefore unable to economically realize many other goals of equal importance. The Society has therefore been unable to aggressively enter the fields of visual education, consumer information, or quality control.

One cannot justifiably criticize the Society for doing what it is funded to do, but it seems to me that if the Society were supported with more freedom in its expenditures, the Society could become a much more vital link in the chain from design to marketing. Why should a respected Society, with years of maturity behind it, be used only as a public relations bureau? Isn't it possible that the Society has a potential function many times its present use?

If there are to be no alternative uses of the Society, and if it is to remain as basically a foreign exhibition bureau, then I would like to encourage the exhibition committee toward expressing the non-traditional, rather than the traditional. Foreign competition and foreign competence have skyrocketed since the early days of the 1950's. When an outside environment is already exploding with non-traditional imagery, it no longer works, for example, to exhibit the same traditional ryijy rugs. This is not to say that the exhibition committee only exhibits ryijy rugs — it is simply an illustration to indicate that a better balance between traditional and non-traditional could be encouraged.

Yes, outsiders do look toward Finland as a leader in form, but these same outsiders are disappointed if they see only a restatement of the past. Finland simply cannot maintain its reputation of leadership when it sets up foreign exhibitions of stylized or dated imagery. These repetitive exhibitions become almost a reversal of image building. If you disappoint your foreign admirers too often, they will soon change their attitudes about you. Again, I do not mean to suggest that the Society has not conducted non-traditional exhibitions — quite the contrary, but what I think happens is that too much of the annual budget is spent on one or two spectacular exhibitions leaving the rest of the year short of funds and with no choice but to mail off exhibitions of ryijy rugs, woven wall hangings and whatever else is cheap to organize. Perhaps a bit more equal distribution of funds over the whole year would upgrade the entire 12 month experience.

Another way for the Society to make its exhibitions a bit more alive, and less monopolized by the present designer-clique, would be for the exhibition committee to gamble on the imagery and the energy of the designers under 30 years old. By traditional standards of general public success, many of these younger people remain untried and untested. If they are forced to wait for exhibition opportunities until they have become members of the acceptable clique, many of them will have abandoned the experimental in favor of playing it safe. This youthful energy needs exposure exactly at the point when it has reached its greatest enthusiasm. It cannot wait around indefinitely — it must expose itself when it is vibrantly alive.

On the other hand, Finland needs to express this energy to the eyes of potential foreign markets. Exhibiting the unexpected imagery of this young designer is one sure means of maintaining the center of attention that seems to be one of the functions of the Society.

Let me here try and illustrate the problem of the inside designer-clique, as opposed to the outside, untested young person — by relating an experience that actually happened while I was writing these pages. Two Americans involved in a very large import of Scandinavian design talent to America came to pay me a visit. Their purpose in coming to me was to find new talent — youthful talent expressing imagery in non-traditional form and material. They came to me for names and addresses of Scandinavian designers in the areas of glass, ceramics, wood, plastics, stainless steel, and textiles. They wanted to begin a whole new line of objects making use of younger talent.

Before they arrived at my door, they had been 3 weeks travelling in Finland, and they were frustrated by the experience. They reported that they had run into a designer-clique in Finland — an in-group — which kept them running around in circles within the same clique membership. They were passed from one established designer to another established designer — and then back to the first — until they were both dizzy and confused. The only way they could break themselves out of this tight clique was to discover someone untried and untested by traditional standards. Interestingly enough, and the reason for relating this example, is that the first contact these people made in Finland, and the contact which put them in touch with the designer-clique, had been the Finnish Society Of Crafts And Design.

I believe the experience of these two import buyers is not an exception. I have myself gone through similar experiences in trying to locate non-traditional imagery for publishing purposes. Far too often, our professional societies are the last to learn about the energy changes which exist right outside of their own doors.

If the Society actively begins to expose this youthful energy through exhibitions, it will be one of the first places the young designer will turn to for advice and encouragement. The young designers of Finland should not have to look upon the Society as a fossilized clique of grandfathers. I believe the exhibition committee of the Society is the place to begin this re-vitalized relationship.

Still another means of securing the capacity of the Society to adapt to change would be to somewhat restructure the Society internally. At the present writing, the Society directorship appears to be a career, or life-time appointment. Although this has many advantages, it also puts severe limitations on the Society's ability to respond to change. When the directorship remains in one man's hands throughout his lifetime, the Society has a tendency to become an extension of one individual personality rather than reflecting a public personality. Perhaps it is therefore in the best interests of the Society to re-organize its leadership structure so that no single individual can dominate leadership for more than a 5-year period. This single step of re-organization can go a long way toward securing maximum flexibility.

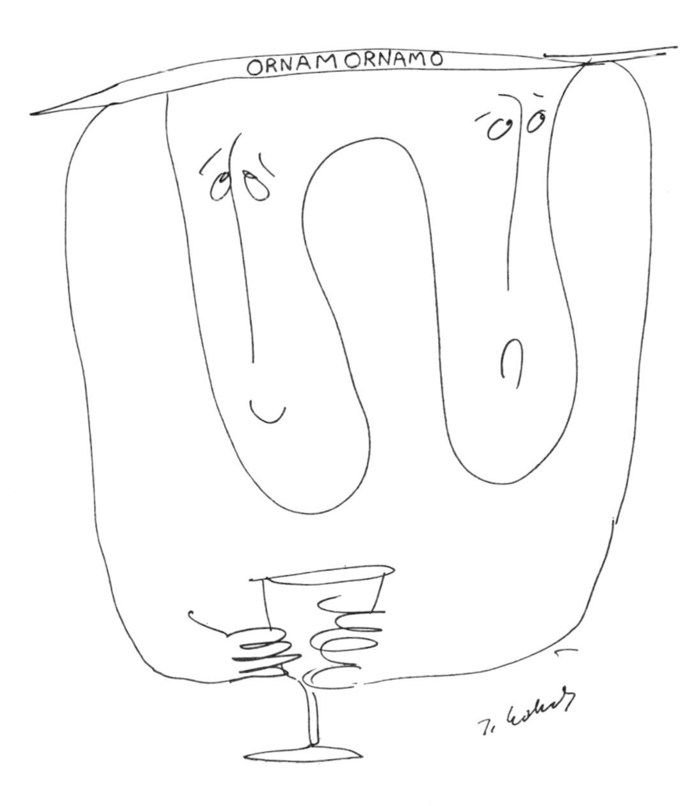

Ornamo

Thus far, throughout the text, I have brought up a number of basic problems, calling them by name — The Communications Gap, The Image Game, The Industrial Grip, The Economic Chasm, The Advertising Myth — and then have added shorter sections on the Ateneum, the Finnish Design Center, Finnhand, and the Finnish Society Of Arts And Crafts. And so now, after coming this long way, we have together arrived at the strongest single point for designer self-help.

There is existing in Finland a designer self-help society which has been sleeping. This self-help society is called Ornamo.

The text thus far has discussed — at times almost too frankly — the environment in which the creative energy of Finland must work. I have called attention to problems and have suggested a number of alternative solutions. For the reader who will accusc me of idealism — let me answer by suggesting that the realization of most of this idealism rests within the voice of Ornamo.

What is Ornamo? Well, theoretically, Ornamo is supposed to be an organization comprising the professional men and women of Finland who are responsible for creating the objects, and most of the object imagery of Finland.

What is Ornamo? Well, up until very recently, Ornamo has been a kind of social club — a club that has been important to belong to, a club that has been important to be seen at, and a name that has been important to use — much in the same way one uses the name of a society country club, or yacht club. Rather than a society with a working, sweating, crying, and shouting voice — Ornamo has until recently been a society with a soothing voice. Instead of a society that has been aggressively protesting against designer abuses — Ornamo has been a safe place to crawl into in order to share mutual condolences.

If there are any doubts about my evaluation of Ornamo's recent past performances, I suggest that the doubting individuals take stock of the long list of concerned designers who have given up their active participation in Ornamo out of frustration. Why have all of these prominent designers given up active participation? Is it, as many doubting voices will suggest, only because these people have made it and no longer need Ornamo? Or, is it because many of these idealistic people decided it was more constructive to try and protect themselves individually rather than to try and cure the collective sleeping sickness within Ornamo? How often has Ornamo been dancing, laughing, and boozing itself to sleep when it could have been protecting designers with advice on contracts, royalties, legal assistance, and dozens of other designer inequities? How often has Ornamo splintered itself into many quarreling sub-groups and sub-committees when it could have been collectively directing its strength into an offensive pressure group?

As of this writing, there appears to be a long overdue rebirth going on within Ornamo. This is indeed a positive sign, and one that I hope will signal a return of active support from the many disillusioned designers who simply gave up.

I suggest that at least part of Ornamo's problem has been the result of its self-splintered organizational structure. Ornamo has divided itself internally into several special branches. There is an ancient axiom that proclaims that "a house divided against itself will fall". Perhaps within this simple axiom lies the key to solving many problems. If Ornamo continues as an identity title for many divergent sub-groups, it weakens itself internally. I suggest that the overall goals of Ornamo are much more important than the special goals of numerous subgroups. Is it not feasibly possible to restructure Ornamo internally into a single-minded entity?

And what can Ornamo do? Ornamo can begin the energy to solve the many problems represented in this book. Ornamo can change its soothing voice into a strong voice of professional dignity.

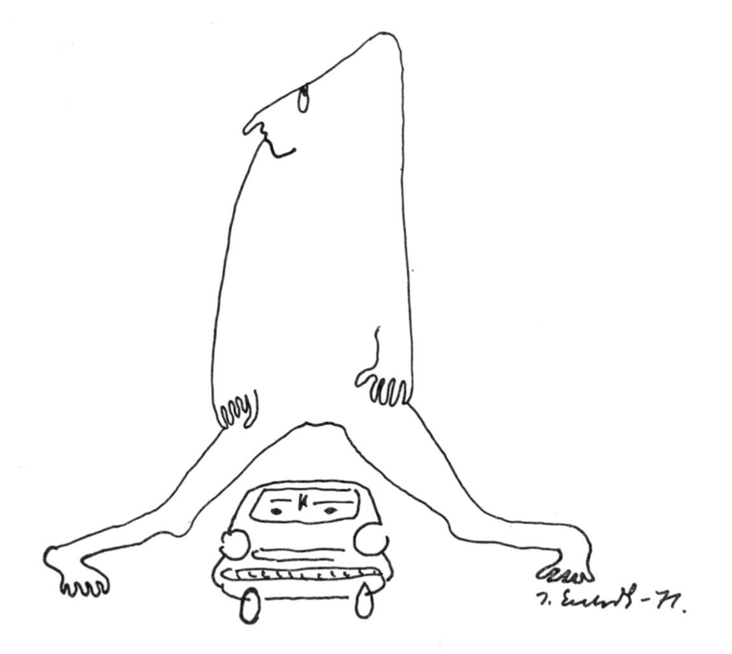

Chapter 10
Postscript

Import-Export

Let me here briefly touch upon two particular Finnish problems of import-export — and by import-export I do not mean objects; I mean the import-export of creative energy.

From what I have observed, it appears that many leaders in the field of visual education believe that the way to broaden the horizons of the Finnish designer, particularly the student, is to import foreigners to Helsinki and have them missionize the natives by telling the natives how it really is on the outside. Quite a number of foreign design people have been imported into this role. The American design prophet Victor Papanek is one good example. Another recent example was the arrival of an American glass designer. There are many other examples. In one sense, I too am an example, although I came on my own.

Although this can be a fine and stimulating gesture, I believe it is the exact opposite of what should be done. For the money that is spent importing foreign spokesmen to Finland to conduct hurried seminars I should like to rather see money spent on regular scholarships to allow Finnish students to study and travel outside Finland. It is one thing to import contemporary prophets from the outside to educate the natives with a speed course — and quite another thing to allow Finnish students to travel and study outside Finland so that they can personally experience what is going on in the hope of learning something useful to bring back home. A guest lecture inside Finland (even though it is a beginning), is not at all the same as a life experience outside of Finland.

Another problem involving import-export is the nasty frequency with which Finnish industry has been importing foreign designers. I have yet to discover something a foreigner can do that a Finn cannot do when it comes to object design. Enough said?

Victor Papanek

Victor Papanek is an intellectual American. He is considered to be a prophet on the subject of design. He is a professional lecturer and an educator. During the past several years he has made repeated visits to Scandinavia with stopovers in most of the major design and handwork schools. He has been in Finland on a number of ocassions, and has participated in seminars at Jyväskylä, and at Suomenlinna. He has also recently written a book which was published in Swedish by Bonniers Forlag in Stockholm.

Normally, I would not have mentioned Victor Papanek in a book about Finland, but his message in Finland has gathered a group of Papanek disciples and also has become the subject of published and spoken controversy. Se here is a new twist — an American gets a special section in a book about Finland.

Mr. Papanek is another design parasite — just like I am — but he travels in academic circles, whereas I try to avoid them. He has published several articles about design in Finland as a result of his visits. I disagree with most of what he has said about Finland and I have voiced my disagreement to him in an exchange of letters. On other subjects relating to design in general I think Mr. Papanek has much to say that is worth saying, but on the subject of Finland, I think his opinions ring with the same sound as that of an American tourist visiting Helsinki on a 24-hour stopover from an around-the-world charter flight. His visits to Finland have often been almost conducted tours — he has been entertained and guided from place to place without the opportunity of going at it alone. His opinions about Finland have come from surface glances, rather than from digging below the surface to see what makes Finland tick. Unless Mr. Papanek takes time to explore Finland on his own, I believe he should frame his comments about Finnish design within the framework of his brief, tourist-like visits.

On subjects not relating specifically to Finland Mr. Papanek offers much wisdom. He has called upon the social consciousness of contemporary Scandinavian designers to involve themselves within the human environment rather than to limit their energies only upon aesthetic proportions of cup handles. He has called upon the designer to look around at the chaotic world he lives in — and has challenged the designer to do something positive about repairing this chaos. I completely agree with Mr. Papanek's challenge, but I do not agree with his proposed solutions.

He has given us a well deserved scolding within the framework of an enormous personal challenge, but far too often he has also provided the specific answers, which to me indicate that he is a far better philosopher than he is a practical man. At the time he first came to Finland he called upon designers to involve themselves with *real* problems, but then he suggested that the *real* problems were, for example, providing refrigerators to underdeveloped people and CP toys to CP (cerebral palsy) children. I suggest that Mr. Papanek was trying to do too many things at the same time. Unfortunately, too many Finnish listeners went right home and began working on refrigerators to save the poor and specialized CP toys to save the handicapped. Mr. Papanek has an important message, but is not always a realistic missionary.

The way to help our fellow man is not to assume that we know the answers. The way to help our fellow man is not to tell him what he *needs*, but to find out what he *wants*. This is the only path of responsible concern. I think Mr. Papanek would completely agree with this. Very often he just gets backed up against a wall by a firing squad of youthful questions — and sometimes his answers look like it. From where we sit in our modern world, we cannot tell underdeveloped people that they need refrigerators dropped to them by helicopter. We must find out what they need by asking them, or living with them long enough to learn. We cannot save CP children from their handicap by designing specialized toys that only remind them loudly of their misfortune. We must ask a CP child what it is that he wants. We might be surprised to learn that a CP child prefers normal toys like a *Barbie* doll, and would be severely insulted by our well-intentioned efforts which only remind him of his handicap.

Another point missing in Mr. Papanek's messages are the small realities of how one goes about realizing an ideal. It is not an easy experience for young ears to hear a powerful message of idealism, and then be hit over the head with this idealism once the realization of the ideal is attempted. This is exactly what happened to several of the Finnish listeners to Mr. Papanek's ideals. He gave them a fine ideal, but he forgot to tell them that they lived in an often cruel and hard business world. The young designer who rushes up to the factory president with a specialized refrigerator to save underdeveloped people gets thrown out on his ear and looked upon as if he were some kind of walnut. I would call upon Mr. Papanek and his Finnish disciples to temper reality with idealism.

A close Finnish friend of mine tickled my insides when he compared Mr. Papanek's visits to Finland to those of the circuit-riding gospel preacher who travels from town to town delivering the same Sunday sermon, and who sometimes forgets where he is, and delivers the same message a second time. I would hope that the next time Mr. Papanek comes to Finland he will turn his record over to the other side — because I know there is another side.

About The Photographs

The photographs in this book do not claim to be exotic, nor do they claim to be "aesthetic". They seek only to be honest.

These photographs are not promotional visuals. Since this book is not a deliberate advertising campaign for Finnish objects, the photographer was therefore under no economic obligation to misuse his camera to improve upon, change, or otherwise glamorize what he saw. Flickering candle lights, frosty imported wine bottles, tables heaped with culinary treats, fashion mannequins, and the normal use of optical and graphic special effects therefore do not appear. If honesty looks a bit dull on these pages, perhaps this is because reality is not a paid advertisement.

More than likely, you will be unable to find many of the more well-known Finnish aesthetic objects, or even the celebrated works of many famous Finnish designers. I hope this is not too much of a disappointment. The book is not intended as a who's-who of Finland. The selection of objects was not based upon a popularity contest, and had nothing whatsoever to do with "fashion", "art", or the status which the object had otherwise received. The objects herein are simply being evaluated — they are not being recommended as status symbols.

When one attempts an overview evaluation of an object-producing nation, then all areas of object production are fair game for comment. For the most part, I have avoided purely decorative objects in favor of the more common functional forms — and even many forms which are not normally considered "design" objects. I have taken a cross-section of objects — from household utensils — to furniture — to problems of packaging — to advertising — to even problems of planning — in the hope of digging into the non-decorative forms which the everyman of Finland must confront in his daily life. For the most part, I have tried to explore objects which, at least upon first glance, seem to either provide alternatives, multiple use, or common sense solutions to some of the requirements of daily life.

As far as was humanly possible, the objects illustrated have all been tested especially for this book. They were not tested by expert testing bureaus — they were simply tested by the people who are supposed to buy and use them. In many cases, objects were borrowed from the firms producing them and they were then placed in Finnish homes, which were only asked that they use and comment upon them.

When an object represents itself as functional, then it seems reasonable to assume that it is useful. One would assume that all objects offered for sale had been thoroughly tested and controlled prior to marketing in order to protect the public interest. Needless to say, our own private, non-expert testing very often brought unexpected results. The man on the street, this Mr. Average Finland, may not be a sophisticated aesthetician, but he is nevertheless quite a competent judge of whether or not the objects he is supposed to buy are well conceived. I am inclined to trust his judgment.

For the most part, there have been no special effects or commercialized gimcracks used in photography. Most of the so-called "design" books on the market take objects out of their natural environment, place them in photography studios, and illustrate them under special conditions in order to make them look much better than they are.

Since the objects in this book are not normally found, or even used in photographic studios, they were therefore photographed exactly where we found them — in somebody's kitchen, in a restaurant, in a shop, or in a public building. The people seen using the objects are not hired mannequins — they are the people who were using the objects either in daily life, or the people we had asked to try them out. These people were not selected according to their good looks, and the women were not selected according to the size of breast measurement. We have diligently tried not to upset reality.

Most of the photographs are therefore what photographers refer to as "candid". These candid photos were taken with a Nikon F camera equipped with a motor drive. When set at full speed advance, the camera will fire four film frames each second — allowing the photographer to record live action in a logical, unrehearsed sequence — much in the same manner as a movie camera. The reason for using this special motor drive was to insure the spontaneity of what the photographer saw.

Many of the objects are therefore presented in a layout sequence of stop-action still photographs. As often as possible, layout illustrates exactly the action taking place while the object was being used. This approach helped to record how well an object functioned without altering reality.

In situations where live action was not important, the photographer tried to capture the function or concept of the object in separate non-action still photographs. His goal was to use the camera to visualize what it was the object represented itself.

The book contains no color photographs. This is not an accident, and it has nothing to do with lack of talent on the part of the photographer. The absence of color is a rational descision in order to keep the price of the book at a mininum. I hope it is consistent with the theme of the book. For this suggestion, I must give proper credit to Armi Ratia of Marimekko. She gave me a well-deserved verbal lecture, over buckwheat pancakes and maple syrup, suggesting that if this was not a book about decorative status symbols, then it should not be itself priced as if it were a status symbol.

Pictures

Can you find the Finn?

Can you find the imitation?

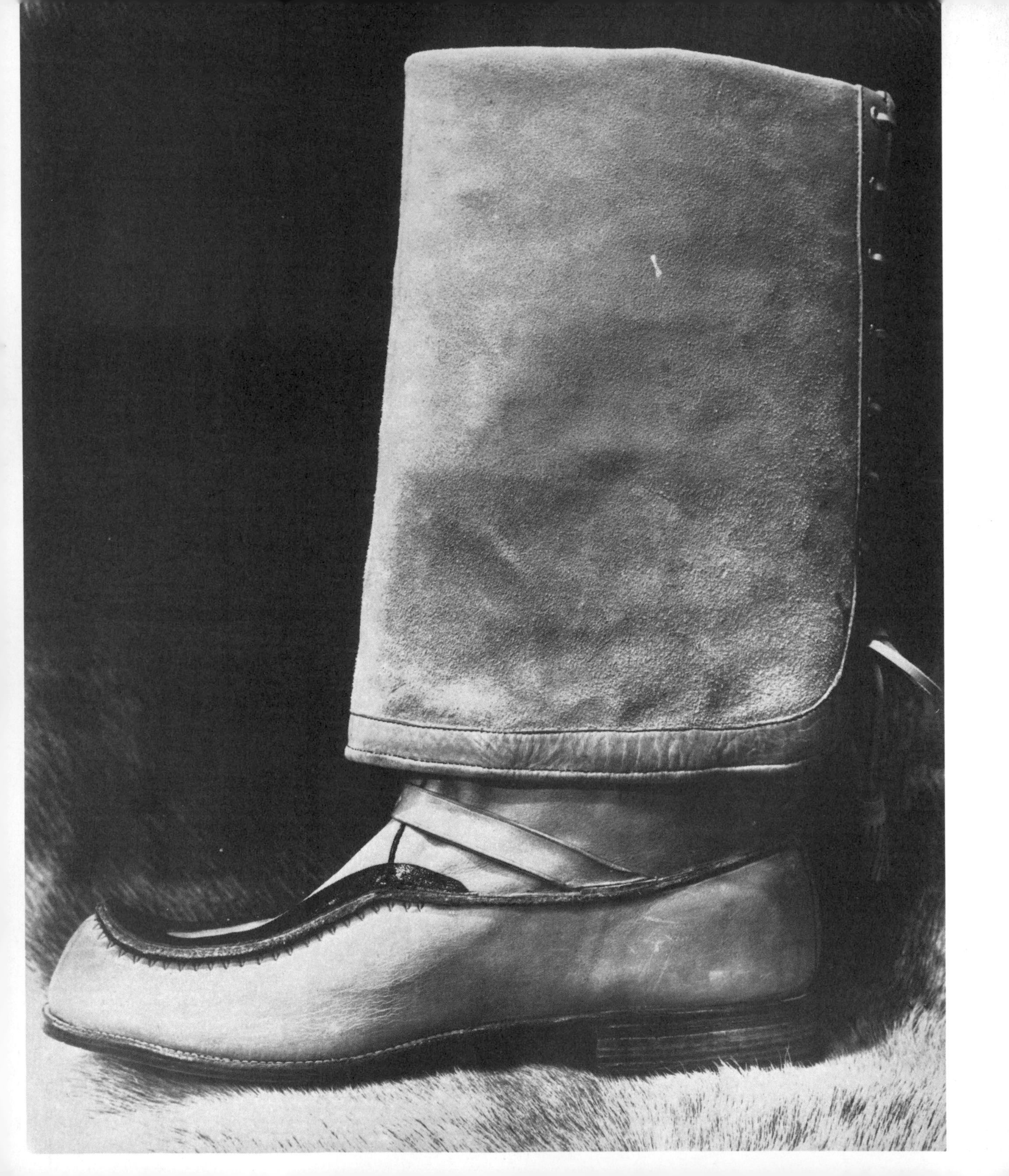

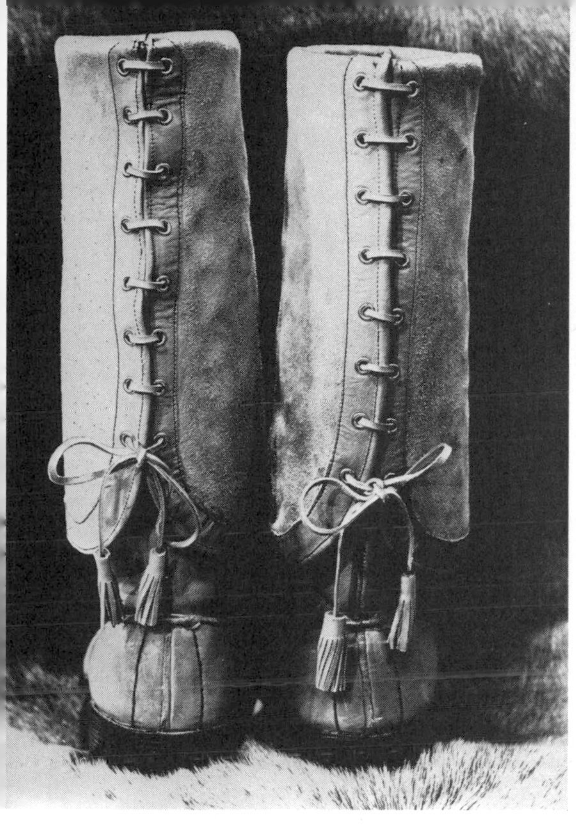

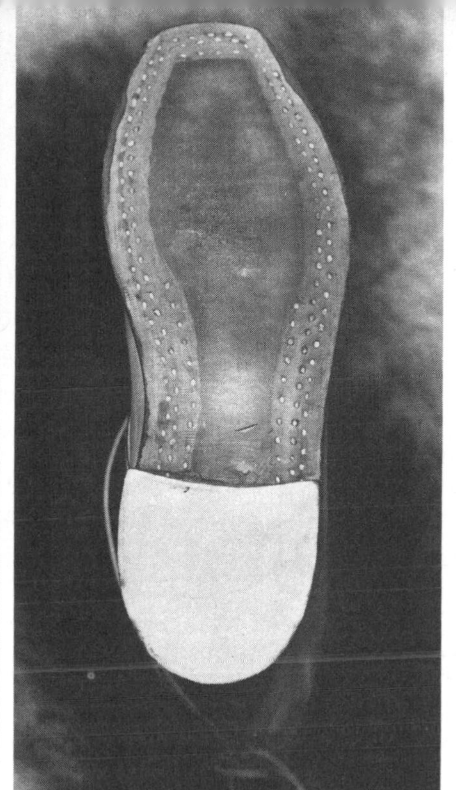

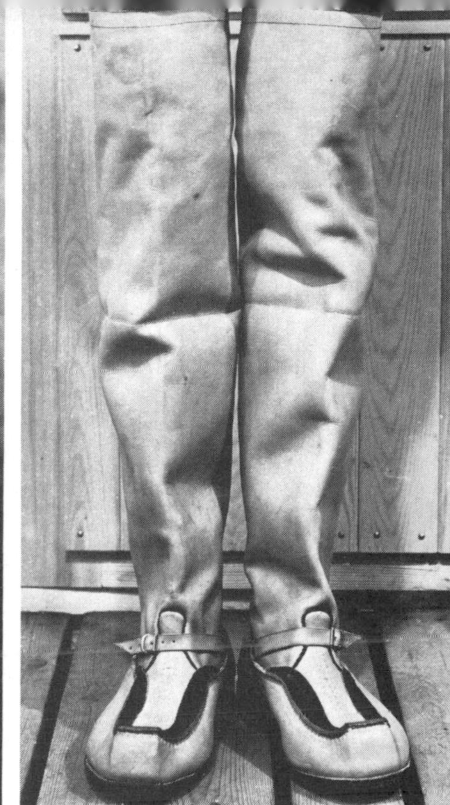

A book which represents itself to cover the subject of Finnish design, along with evaluating general objects, must also make an aggressive attempt to explore those objects which reflect something uniquely Finnish about Finland. To my eyes, this traditional high-low leather boot handsewn by Tuupotuote Oy answers this demand. This boot form, although I have rarely seen it considered among the advertised objects called Finnish "design", has been worn and enjoyed in the past by thousands of Finnish landsmen.

By 1970, and after spending many days looking for this high-low form in Helsinki shoe stores, the form had sadly enough all but disappeared from contemporary production. Tuupotuote Oy of Tuupovaara is one of the few firms keeping this once common Finnish form alive. It is indeed ironical how this practical and beautiful traditional Finnish form has been completely overlooked during the recent high-fashion trend toward both men's and women's boots. I have often wondered if at least symbolically, this boot form marks the beginning of the end to handwork in Finland. I have myself shown this boot to leather craftsmen all over both Scandinavia and the United States. It has received endless compliments.

The boot soles are hand pegged to the uppers with wooden pegs. As illustrated, Tuupotuote Oy has carried the same form idea out in the low boot, the lace shoe, and in a sandal model.

The illustrated pair of boots without laces were photographed in the home of Oili Mäki, and have been used for over 25 years.

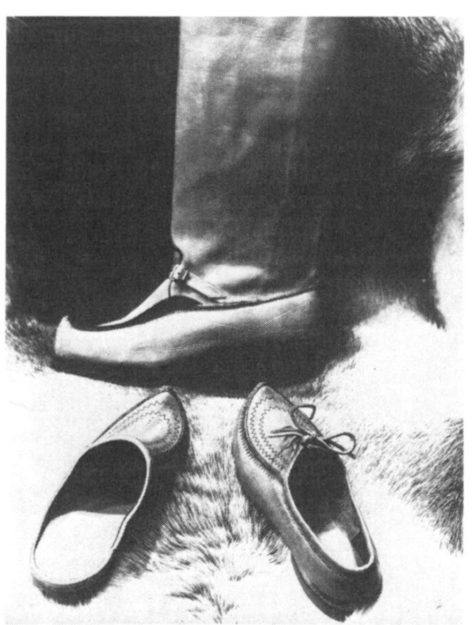

Although felt boots are not exclusively a Finni
tradition, they are manufactured in Finland
by the hundreds upon thousands of pairs.
They represent the survival of a traditional for
which has been used for generations, and which
has been improved upon with the modern additi
of rubber soles and heels. The boots are
inexpensive, seamless, extremely lightweight,
and their comfort and warmth are enjoyed and
appreciated by thousands of Finnish men, wome
and children every winter.

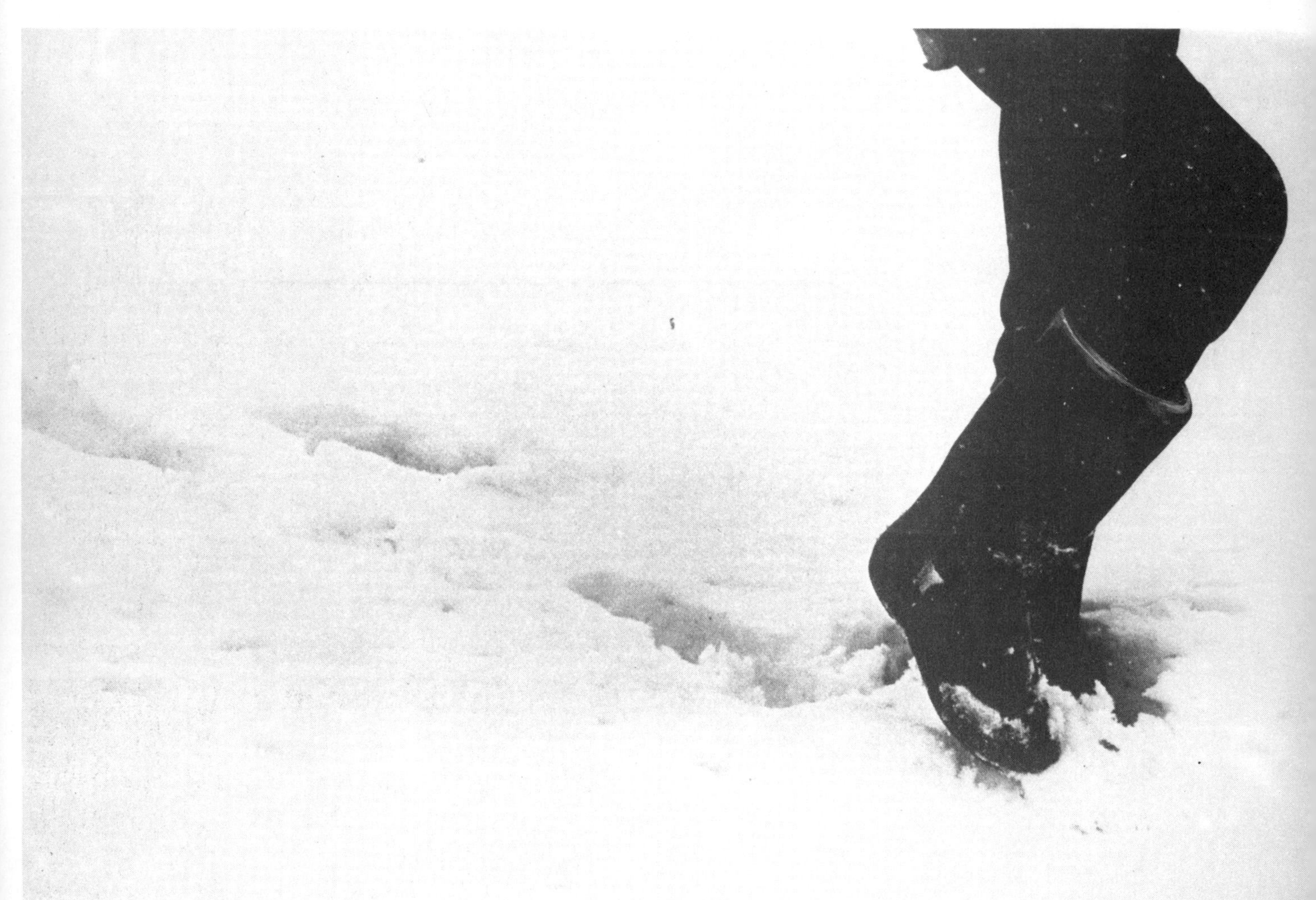

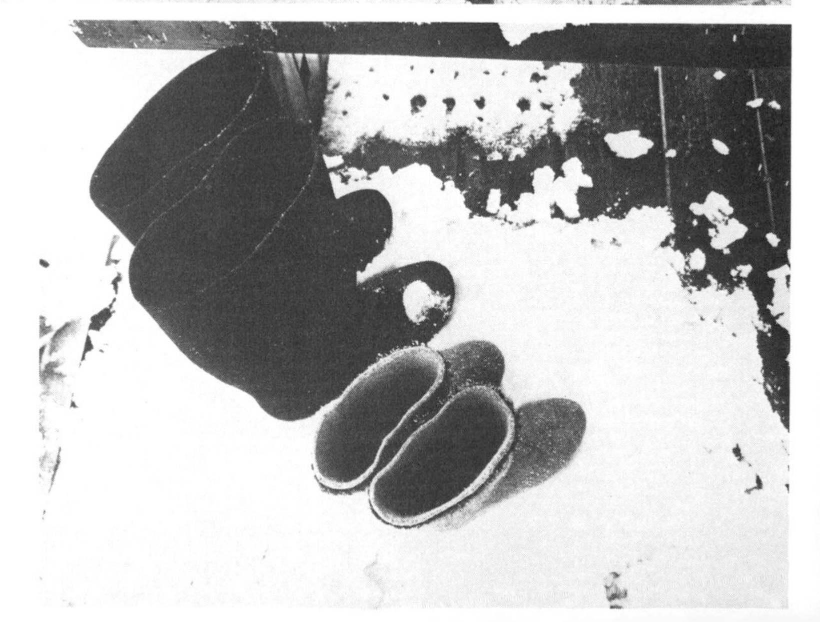

I believe the felt boot,
perhaps because it is another non-intentionally "designed" form, has been one of the most underrated, but yet potentially valuable export forms in Finland. The boots represent a form which has been overlooked, taken for granted, and not fully appreciated by industry for its export value. With a bit more refinement of the form, and with the recent addition of multi-colored felt and the waterproof soles — the form could well become a Finnish contribution to winter footwear around the world. It is ironical that even though these boots are as common to Finland as the sauna — they are only rarely considered among the objects of Finnish "design".

Mixing device, carved from a tree limb. Does this multifunctional tree limb, which one sees in use all over Finland, have something to do with Finnish design? Or, is "Finnish design" a piece of industrial glass which has a label glued on to it reading: "Finnish design"?

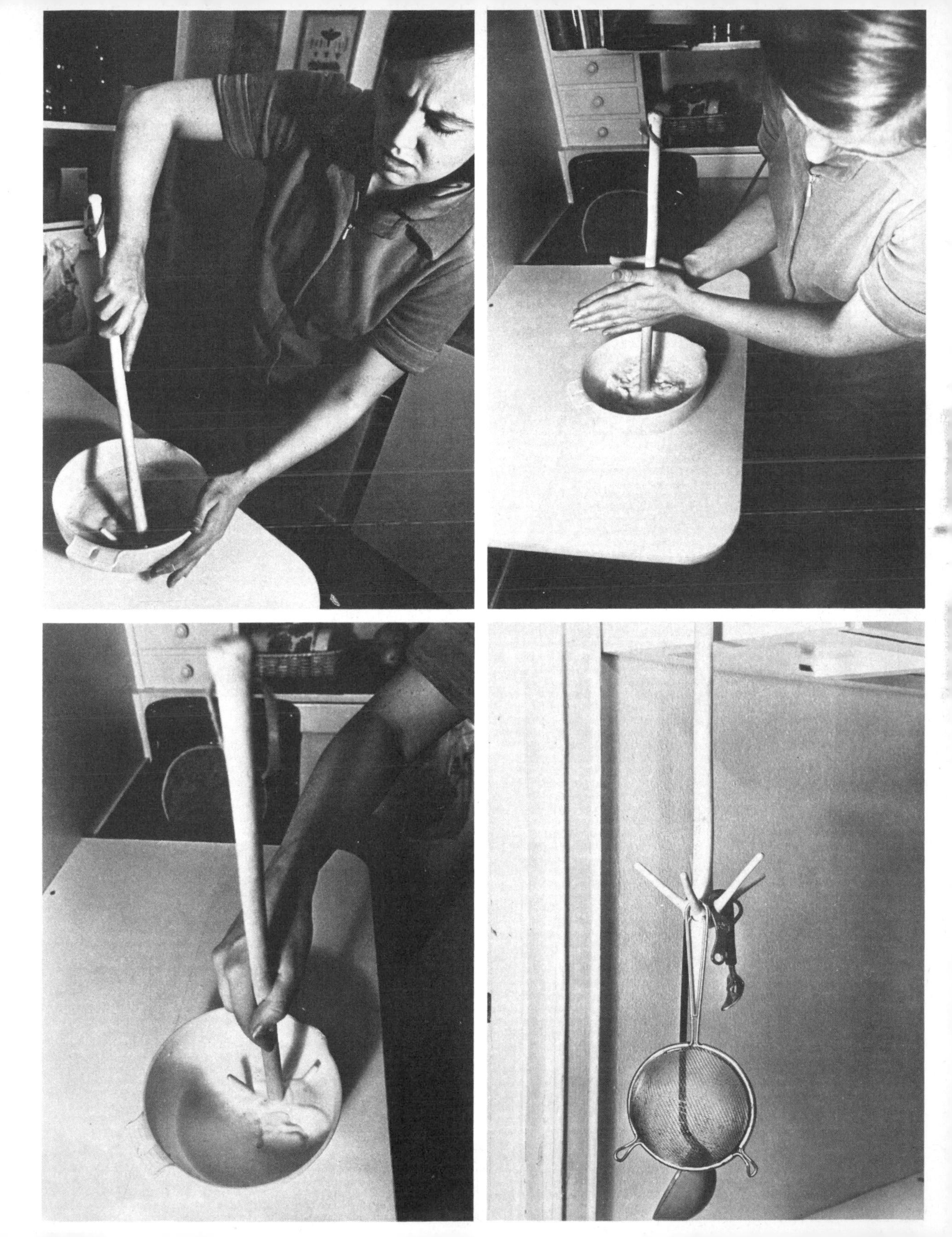

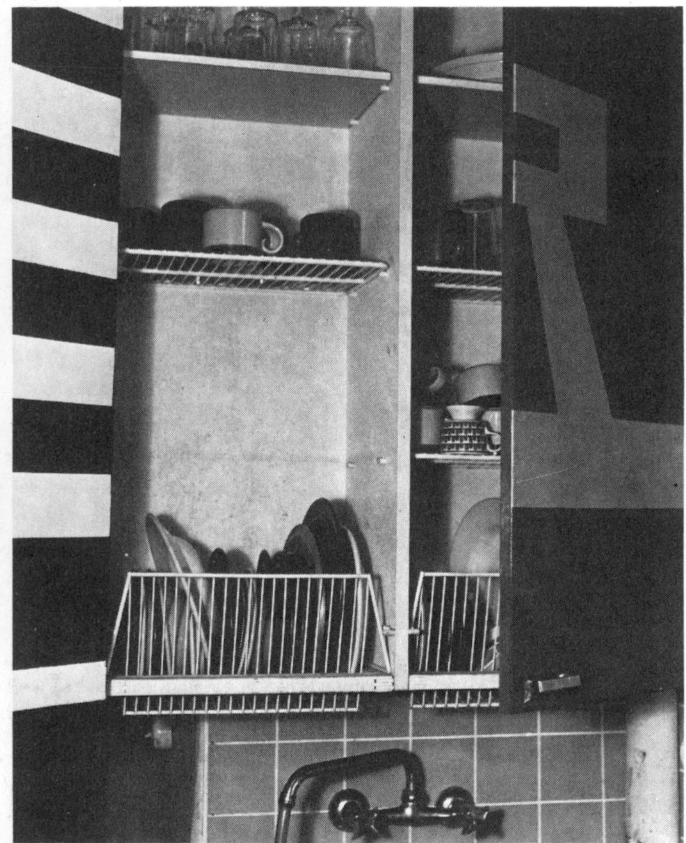

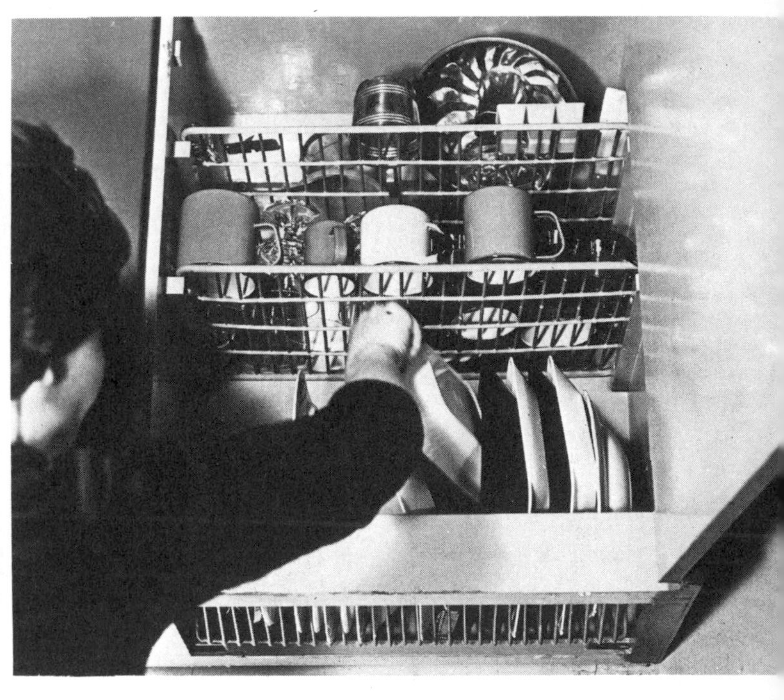

Although I have never once seen ordinary Finnish kitchen cupboards illustrated as Finnish "design", these drying-rack cupboards do appear to be uniquely Finnish. I have seen them nowhere else in Scandinavia, central Europe, or even in America. They are so common in Finland that they are taken completely for granted, but yet they are so basically functional that they make all other kitchen cupboards look incomplete. As the photographs indicate, the cupboards are almost always built directly over the kitchen sink. Rather than wooden shelves, the cupboards consist of drying racks. Wet dishes are placed for permanent storage in the racks, and since the cupboard is open on the bottom, it allows air to circulate into the cupboard to dry the dishes. The automatic dish-dryer of Finland — it uses no electricity, it has no moving parts, and it never needs repair!

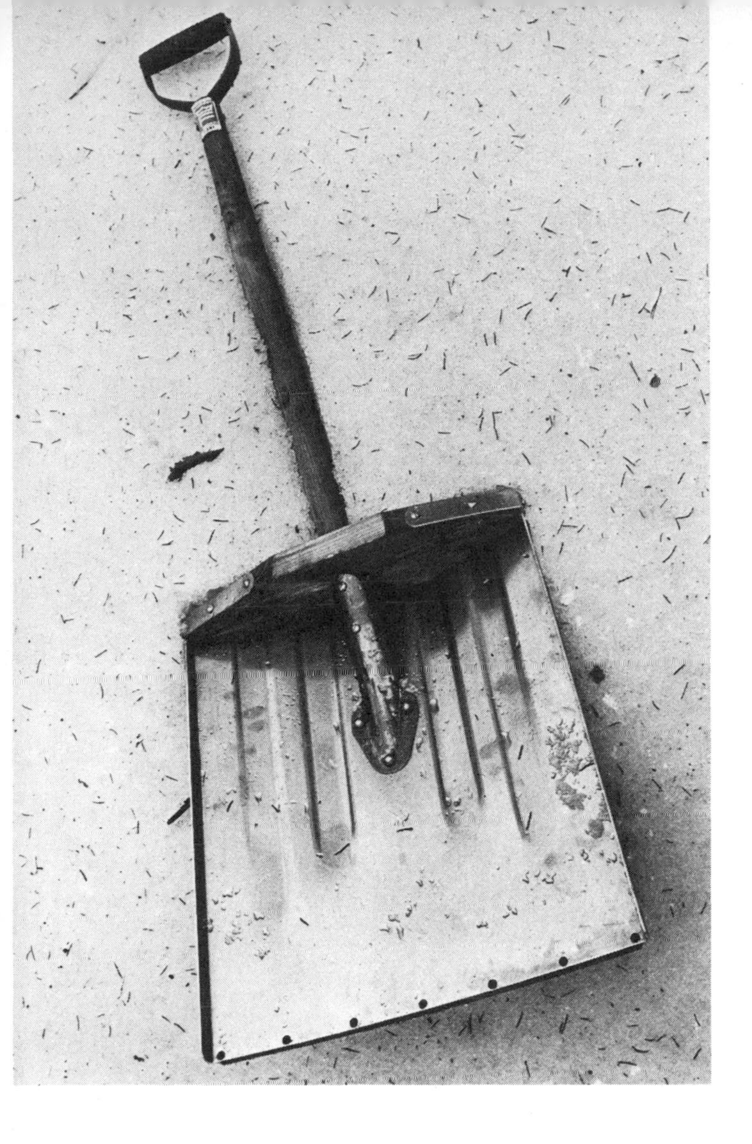

Wood and aluminum snow shovel — a traditional snow shovel form, but with the modern, functional addition of lightweight, durable aluminum. This Fiskars snow shovel does not claim to be "designed" — it simply functions. It seems important to me that Fiskars had the good sense to preserve this timeless form and then improve upon it, rather than to cast it aside in favor of something more "aesthetic". Although almost every home-owner in Finland owns or uses one of these shovels — it is not normally considered among the objects of Finnish "design". Curious!

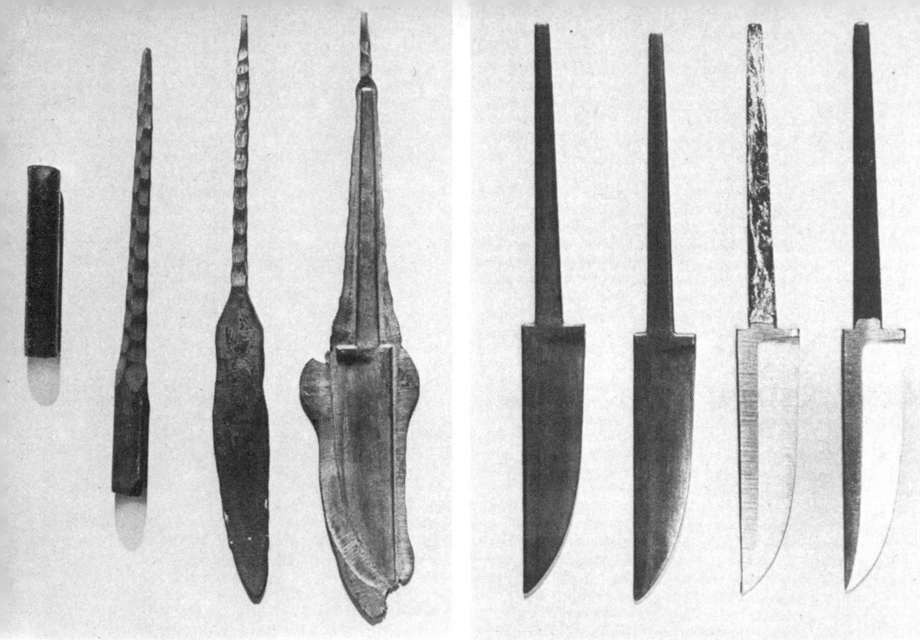

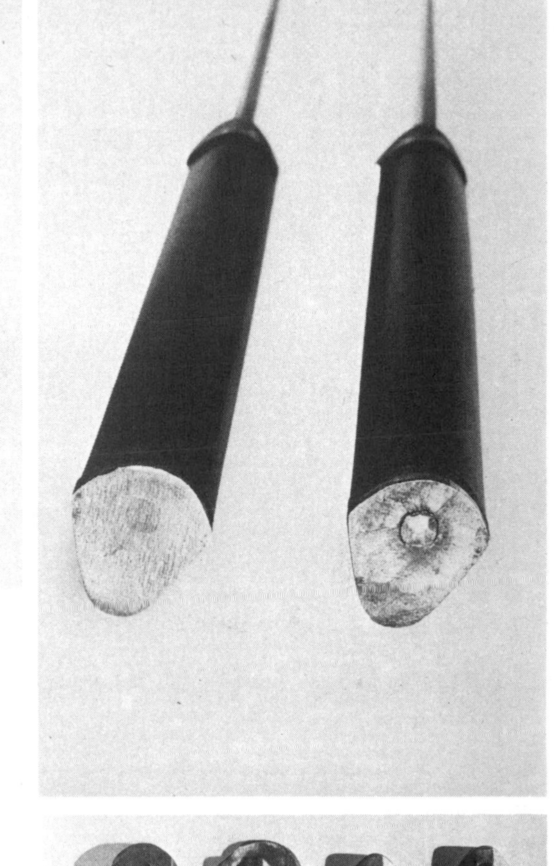

e photographs on these two pages illustrate a comparison between a ditional Finnish "puukko" knife, and a modern adaptation of the ne form. The traditional form as illustrated was made by hand in studio of Johannes Lauri of Rovaniemi. The modern "puukko" s designed by Tapio Wirkkala and is manufactured by Hackman Oy. "puukko" has a long tradition in Finland. Above all, it must ction. The form is used by thousands of landsmen, woodsmen, ermen, and hunters. It is a form which is meant to be used — and d continuously. The Wirkkala form by Hackman Oy has received a at deal of positive critical praise, and many writers feel that it is exceptional example of where a traditional form has been improved on by modern technology. Most critics also feel that Mr. Wirkkala's dern "puukko" is a highly "aesthetic" form. It is my feeling that ch of this positive praise for the Wirkkala "puukko" has come m the pen of critics who have failed to try the knife out before noring it with their words. I do agree that as far as one can define dern "aesthetics", the Wirkkala "puukko" is a beautiful form. But I o feel that it represents a deliberate "design" product, and that it is an mple of a form where industry is using the good name of a signer to market an inferior object. Hackman Oy has prominently nted Mr. Wirkkala's name on the knife blade, and from what I have en able to observe, the knife seems to be marketed as an "object of ", rather than as a functional knife. For the most part, Mr. Wirkkala design a functional knife, but somewhere along the line — haps when it became an "object of art" — the form began to lose functional qualities demanded of the traditional "puukko". The rkkala knife is manufactured in two sizes — 18 centimeters overall gth, and 21 $^1/_2$ centimeters overall length. It has a stainless steel blade h shaft running all the way through a nylon handle. The blade is d to the handle with two brass fittings. Mr. Wirkkala designed the fe with a flat handle end so that it could be used for pounding. This dition is an admitted improvement over the traditional "puukko". . Wirkkala designed this Hackman knife for maximum blade strength running the blade shaft completely through the handle — this is other improvement over the traditional "puukko". He has also created ine harmony in his use of materials — stainless steel, brass, and nylon. t where this modern "puukko" falls completely short of its mark is in the ality of material used in the blade. The Hackman firm gave me several ives to try out and these were all placed in Finnish homes for testing.

The major complaint is that the blade is too thick, will not hold a cutting edge, and is a much better knife for spreading butter than for cutting. As can be seen in the comparative photographs, the blade on the Wirkkala knife is also shorter than the blade found on most traditional "puukko" knives. This short blade, when combined with a quality of stainless that will not hold a sharp cutting edge — reduces the function of the Hackman contribution. In retrospect, I believe that Mr. Wirkkala's form was well conceived on the drawing board, but it leaves much to be desired from a manufacturing standpoint. The Wirkkala form is also easy to wipe clean, but if the blade won't hold a cutting edge, then the whole concept fails in its basic function. It also seems to me that since this knife is intended for outdoor use, an inexpensive sharpening stone could be included with the knife in a side pocket to the sheath. In many ways, this whole knife, although it has been highly praised, seems in conflict with itself. If it is the ultimate "designed" product many critics claim it to be — then why is it packaged in a non-descript plastic bag? Additional photographs indicate the fabrication parts and construction principles involved in the Hackman "puukko"

What is beautiful, highly functional, Finnish, and reasonably priced? How about the wooden doorknobs of Finland? These photographs compare ordinary wooden shop doorknobs with both a modern acrylic doorknob in the traditional form, and the cast brass doorknobs of Alvar Aalto on the new Helsinki Stockmann bookstore. Mr. Aalto's doorknobs are architecturally "aesthetic", and they are also ultra-functional, and ultra expensive. Perhaps a comparison between the traditional wooden doorknob and the Aalto doorknob of Stockmann's symbolically represents where Finland has been, and where Finland wants to go?

TYÖNNÄ
SKJUT

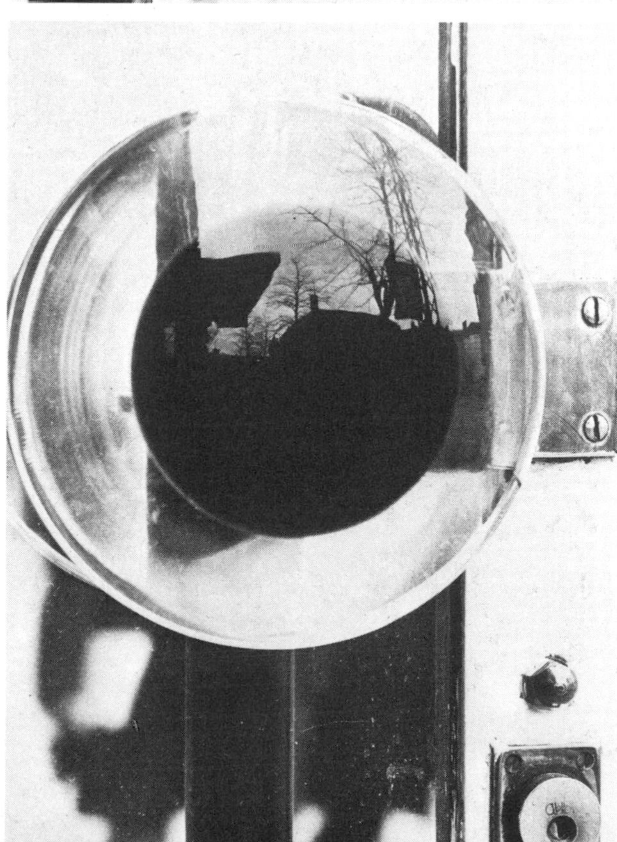

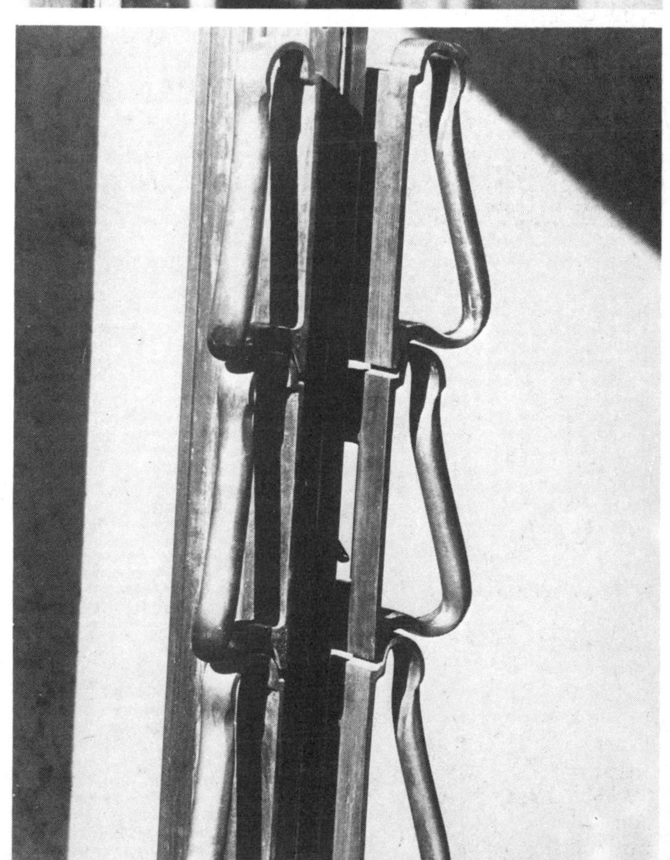

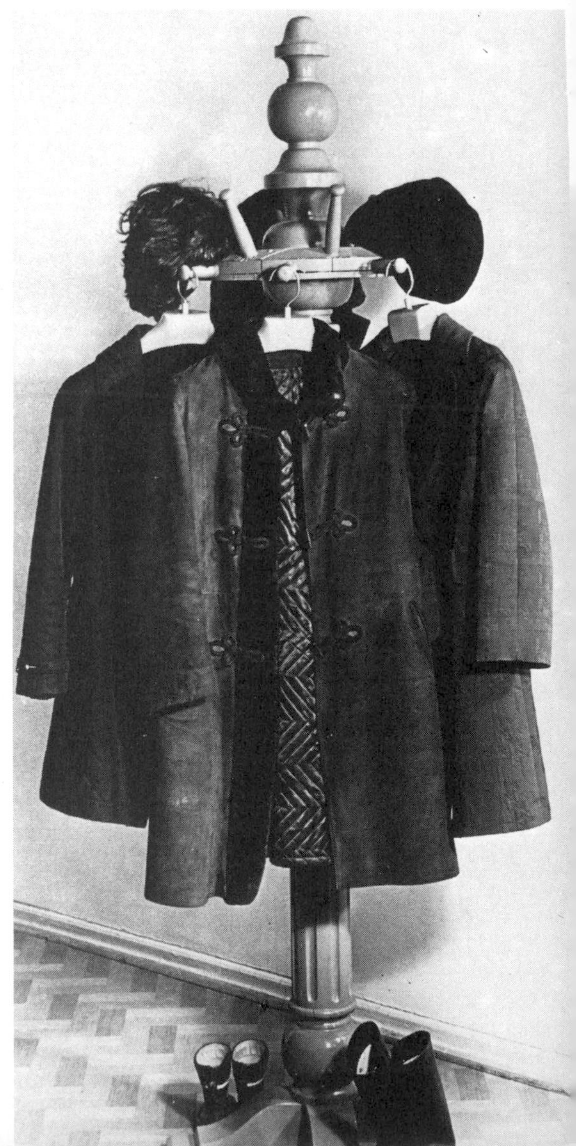

Plastic coathangers by Sarvis — a form which seems to defy improvement. It represents a combination of function, responsible pricing, and beauty at its maximum point.

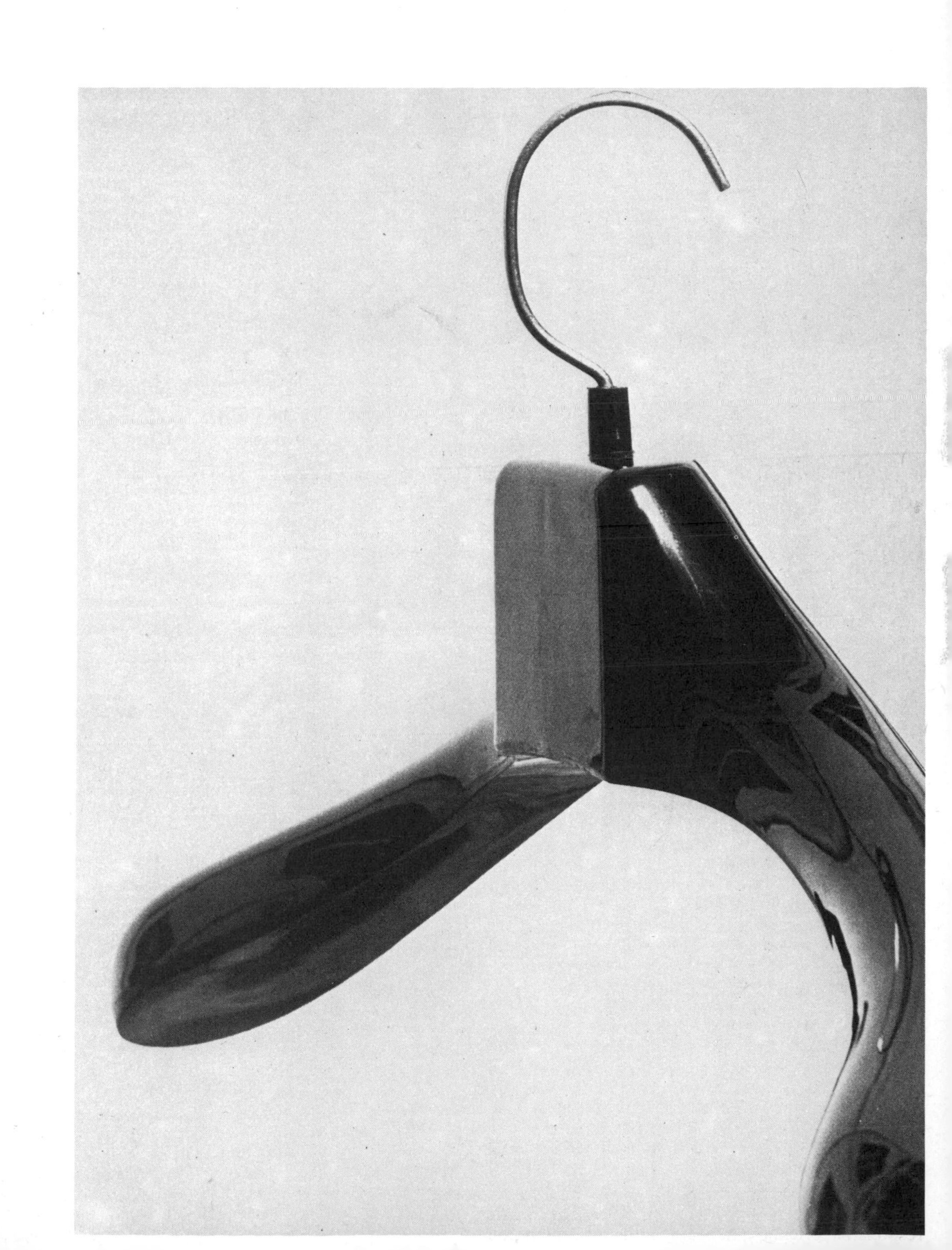

"Pilkkipulkka" shelter sled by Fiskars. This form does not represent itself as "designed", but yet it seems to be what rational design is all about. It is one of the everyday, non deliberately "aesthetic" objects of Finland. The multiple use form is made out of aluminum scrap which is leftover after the fabrication of large Fiskars boats. It can be pulled or tied from either end, and it is strong enough in its construction to withstand the weight of a grown man, two children, or even a full load of fishing gear. The form was developed particularly for ice-fishermen, but can also be used as a child's sled, or as a vehicle for hauling. When not in use as a sled, the form becomes a functional wind shelter with the added convenience of a small storage compartment.
On the day we photographed the "Pilkkipulkka", some of the snow on top of the ice had melted into puddles. As can be seen from the sequence of photographs, the sled proved its strength with the unexpected, and emergency rescue of a stranded Fiskars employee who had come with us into the countryside, but without his boots.
The ice drill used by the fisherman is another example of non-"designed" design, also by Fiskars.
On the way in from the lake, and after finding no fish, the fisherman found a new use for the form as a portable lunch table in the snow.

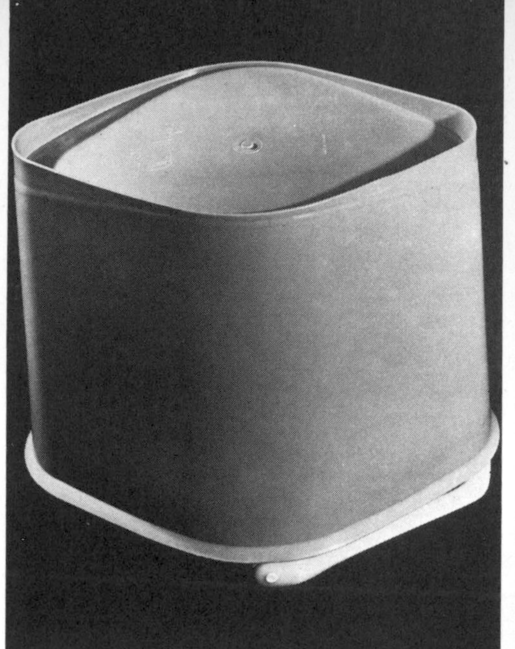

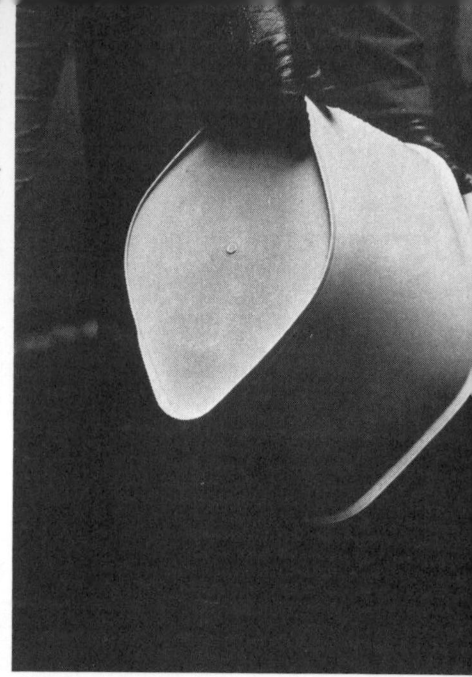

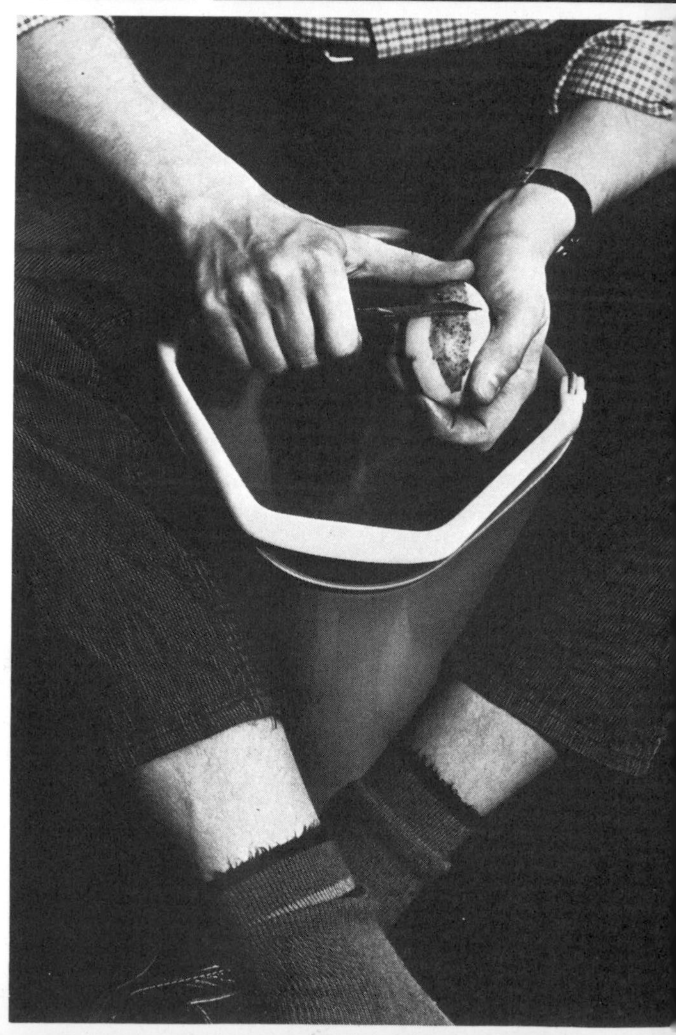

Plastic, multi-functional pail by Olavi Arjas, and manufactured by Sarvis. Mr. Arjas who is an engineer, rather than a designer, related an interesting anecdote in explaining the background for the unusual form of this pail. Mr. Arjas has a weekend house in the countryside. During the winter, when the water pipes are frozen, he has had to regularly carry water from his well by the pail full in order to fill up the water tank on his sauna stove. For many winters, he had been carrying the water in an ordinary round water pail, and as all of us know only too well, the round pail will not hang next to the body. Mr. Arjas grew very tired of having ice cold water from a round pail splash over onto his trousers, and so he decided to do something positive about the problem. This Sarvis pail was the result. It can not only be carried closer to the body, but the handle is shaped for a firm, non-slip grip. The form is deliberately shaped for pouring, and the overhanging sides create a lip along the bottom to hold onto when pouring. As an added practical matter, the pail also fits snuggly between the legs when peeling potatoes, or picking berries.

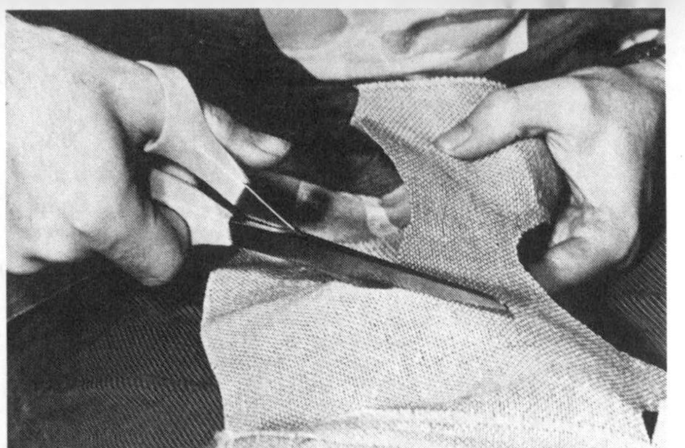

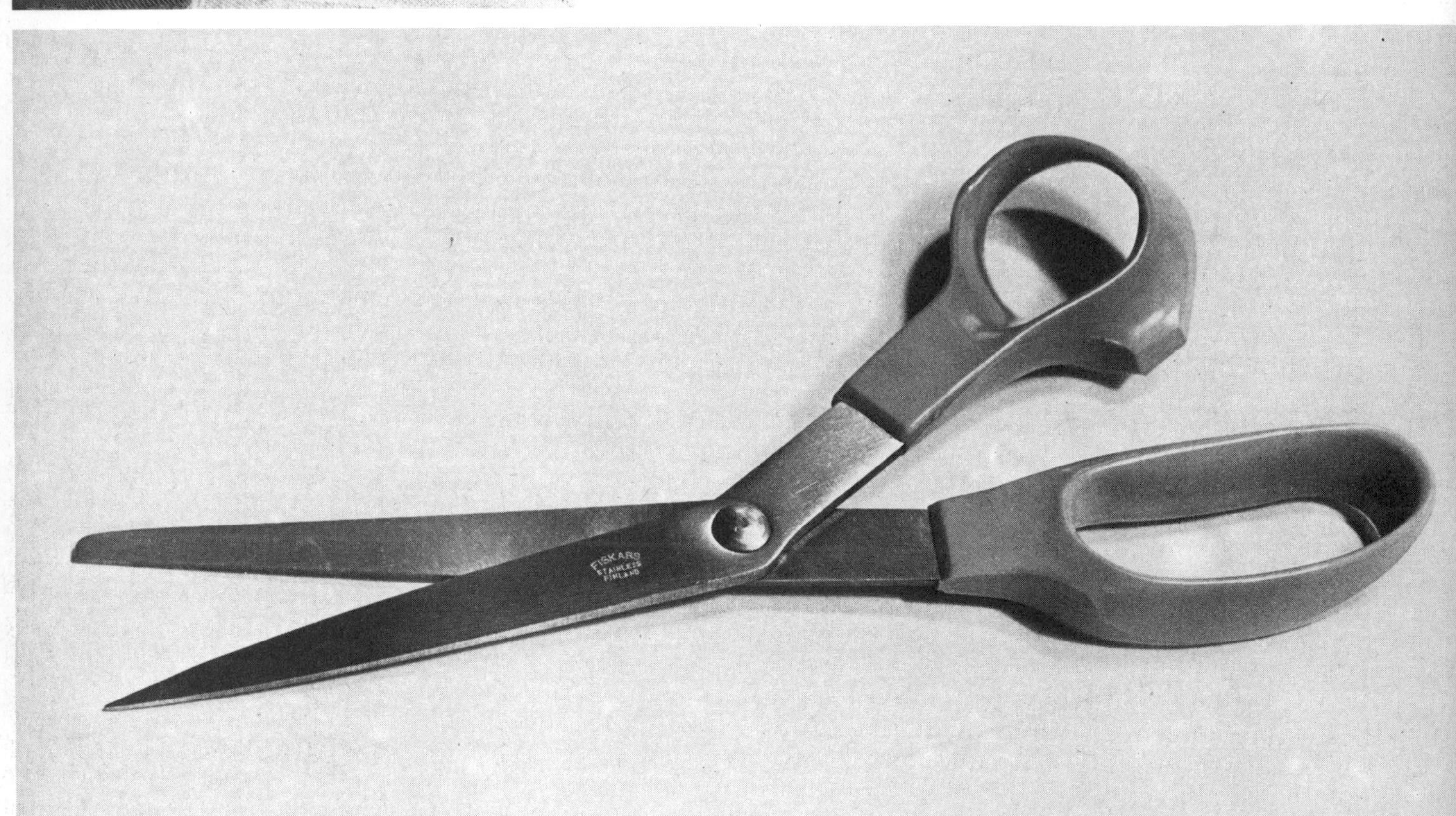

Fiskars scissors — stainless steel blade with colorful plastic handle. Without reservation, and as far as I have been able to compare, I believe this scissors is the best household scissors manufactured anywhere in the world. The colored handle is easy to notice, convenient to hold, and beautiful to look at. The permanent blade construction and cutting edges make this scissors ideal for use in cooking, cutting textile, cutting paper, and even cutting up to 2 millimeter thick leather. This common household form represents design at its most rational point of perfection.

Handleless plastic cups by Sarvis. This form does not represent itself as "intentionally designed", but yet it out-functions most plastic cups and even many of the ceramic cups on the Finnish market. One Sarvis cup sells retail for well under 1 Finnmark. They are produced in a variety of bright colors, they are more resistant to tea stain than many ceramic cups, they conveniently stack, and even without a handle, one can hold a cup filled with coffee in the bare hand. When compared with its major Finnish competitor, this Sarvis cup with no handle and an excellent quality of plastic, is several times more functional. The competitive plastic cup is advertised as "aesthetic", but the quality of the plastic is not as good as on the Sarvis cup, and the handles of the competitive cup are so ill-formed they are next to useless.

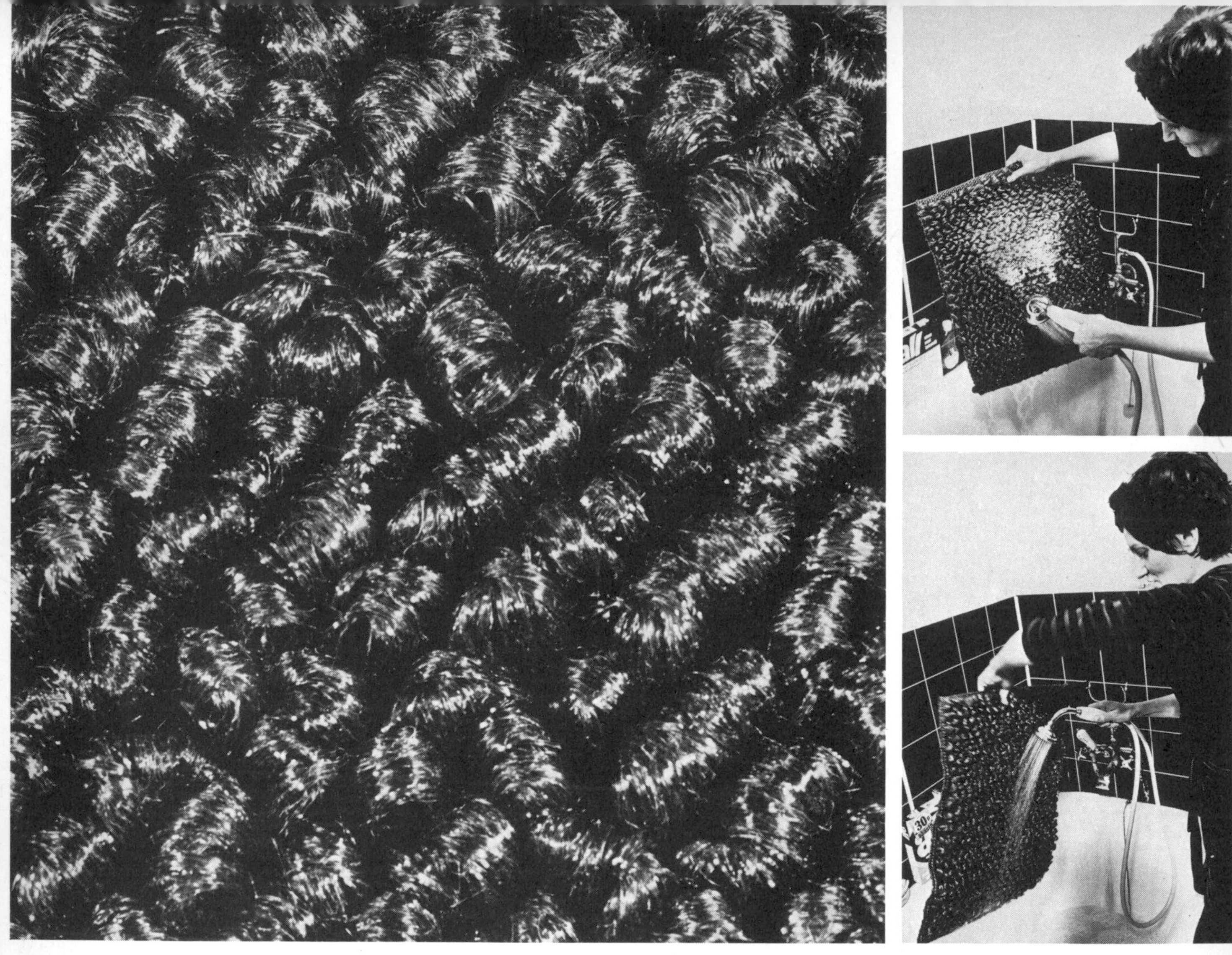

"Bioliitti", bathroom, or wet room carpet by Metsovaara Oy. This carpet represents itself to be almost indestructible and is made out of 100 % polypropen fiber. The fibers will not absorb water, and are highly resistant to temperature change, rot, oils, and acids. The carpet is supposed to answer the practical demands of the modern housewife because it can be washed quickly and conveniently by simply dunking it in the bathtub or shower. A Finnish housewife tested this carpet and endorsed its durability, but commented that to stand on it with bare feet is like standing in a field of hay stubble after the hay had been cut. Metsovaara's main problem in using the fiber is that they weave it by hand. This places the carpet beyond the economic reach of the average housewife who is supposed to benefit from its durability and convenience.

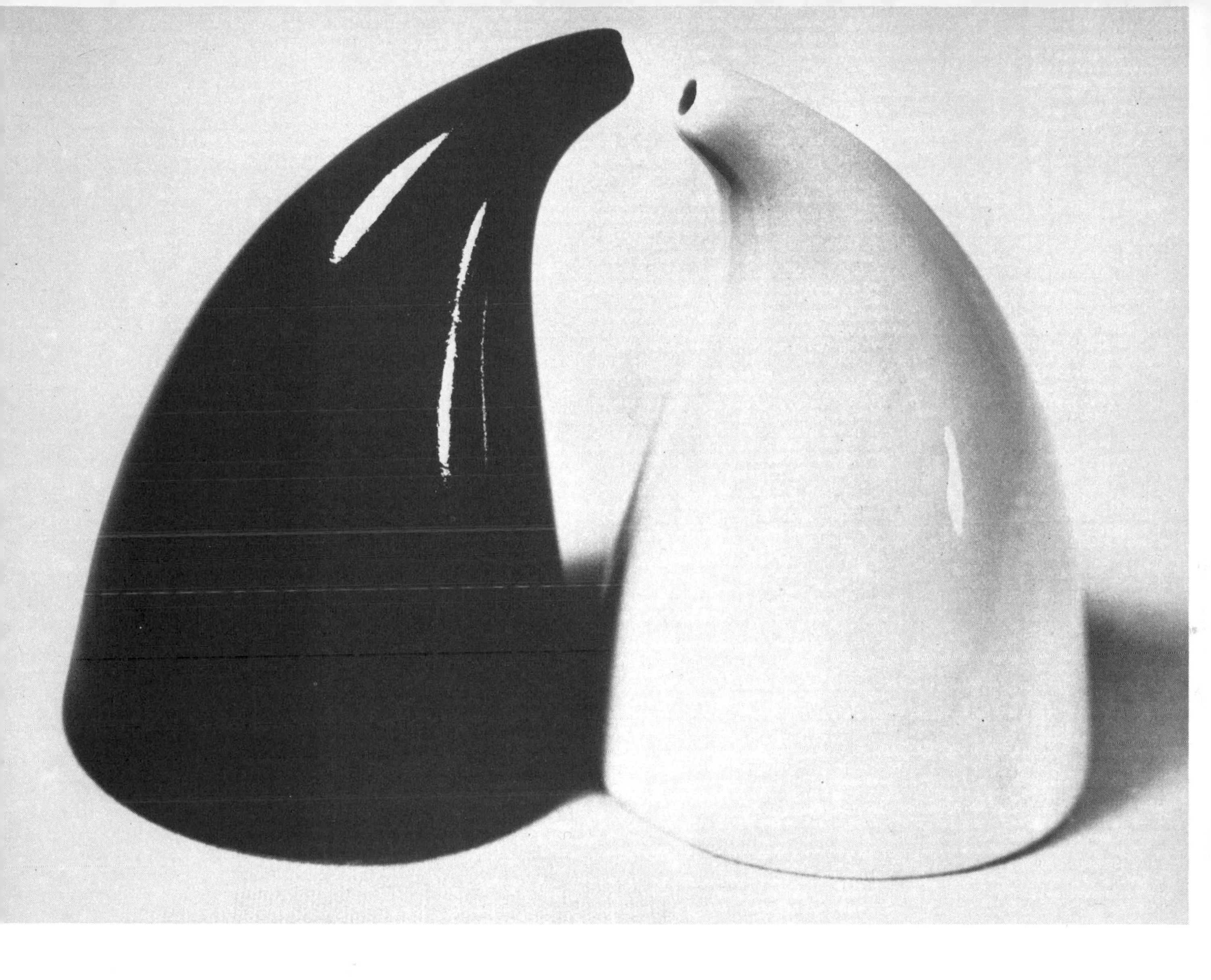

Teardrop salt and pepper set by Kaj Franck, and manufactured by Arabia. The form, the size, and the contrasting colors in the set make it not only functional, but also beautiful. The only problem encountered with the form was an occasional incident where either too much pepper flows, or the salt (especially in summer) gets clogged up in the one hole. Is it possible for "design" to come up with a salt and pepper set that gives the consumer functional alternatives for setting the desired size of opening?

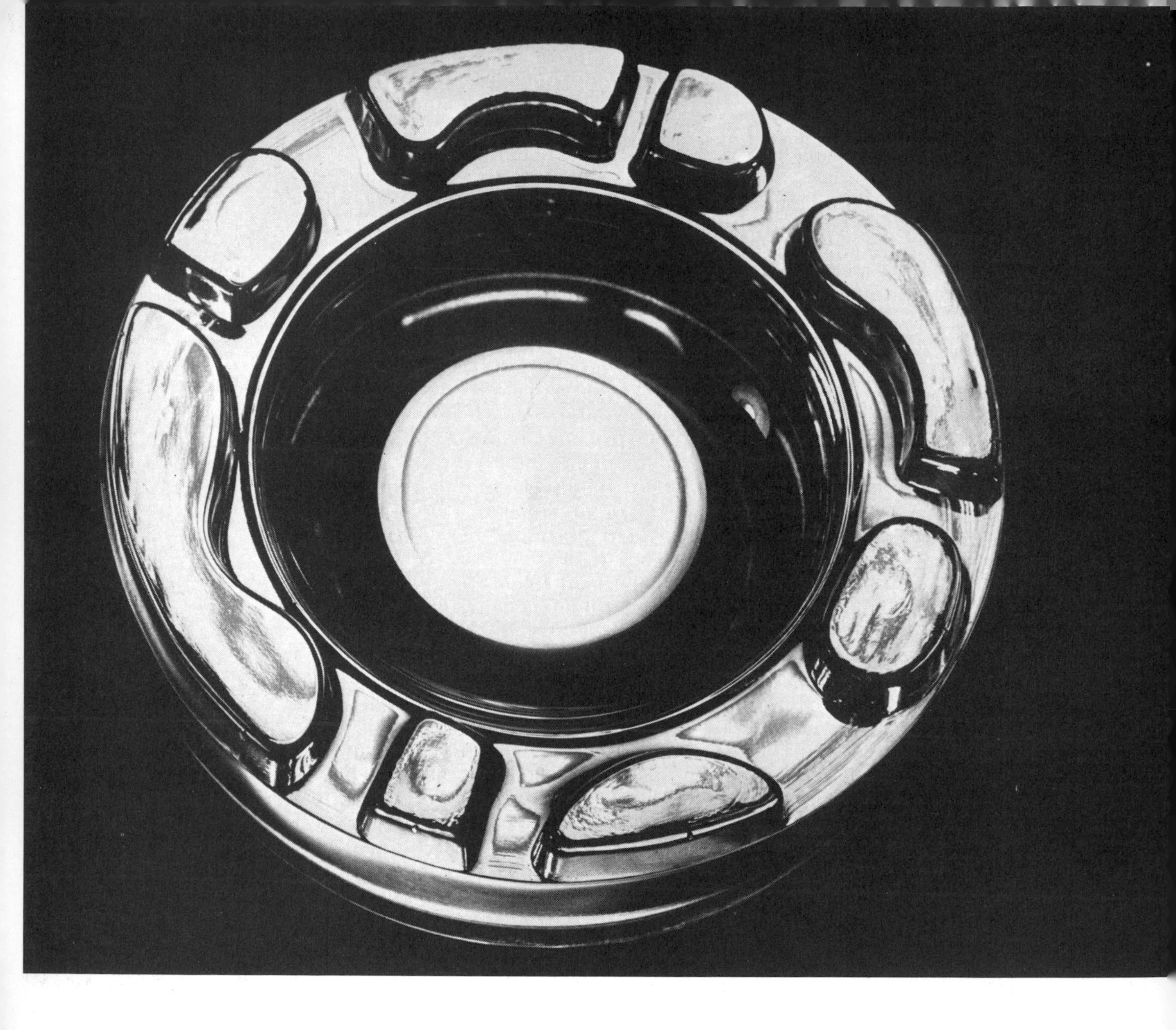

Multiple use glass ashtray by Heikki Orvola, and manufactured by Nuutajärvi. While in the Arabia-Nuutajärvi showroom I found this ashtray, as if overlooked for its multi-functional value. At the time these photographs were made, the form was being manufactured only in non-colored glass, and it was inexpensively priced. Mr. Orvola obviously was keenly aware that modern tobacco manufacturing had produced cigarettes and cigars in a variery of thicknesses and lengths which no longer fit the standardized ashtray forms of ten years ago. His abstract surface pattern on this circular form tightly holds everything from English ovals to mini-cigars. Perhaps this may seem like a very small nuance in the overall picture of design, but nevertheless it marks a responsible convenience for smokers.

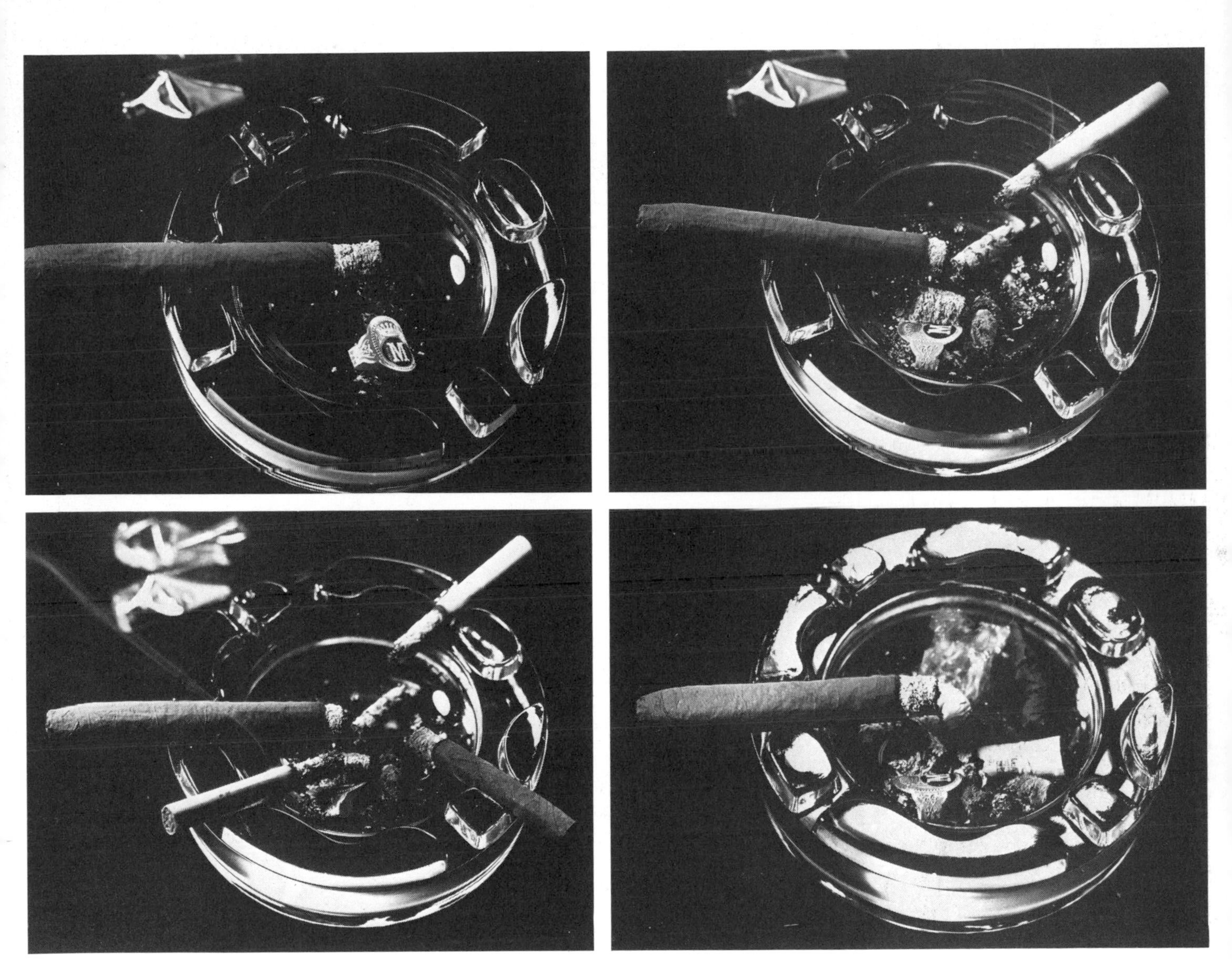

ketchup

An almost modern antique ketchup bottle in plastic by Tapio Wirkkala that still continues to work, and work, and work. It hardly ever misses, and it rarely ever drips. It just works. Mr. Wirkkala designed this form for use by Strömfors and Huhtamaki — a workable combination of ketchup and plastic.

"Helmi" glasses by Heikki Orvola and manufactured by Nuutajärvi. This glass form answers all the demands of both "aesthetics" and function. As can be seen from the photograph, Nuutajärvi has also taken the additional trouble to package the glasses in a functional and attractive display box.

Coffee cup, saucer, and cover — part of the "Ruska" series by Arabia. To make this photograph, we borrowed a cover from the "Ruska" series sugar bowl, which incidentally happens to fit the coffee cup, but is too top heavy to be practical. Why not have coffee cups with optional, low cost, self-sealing covers for all of the thousands upon thousands of morning and afternoon coffee drinkers who work, and who appreciate a hot cup of coffee, rather than one that is only lukewarm? Arabia provides this cover convenience on institutional cups, particularly for hospitals, but not for the general public. Since it is not uncommon to find covers on throw-away coffee cups, why is it therefore hard to find them on permanent cups? Although this "Ruska" set is inexpensively priced, we otherwise found the saucer rim too narrow to conveniently hold most normal teaspoons.

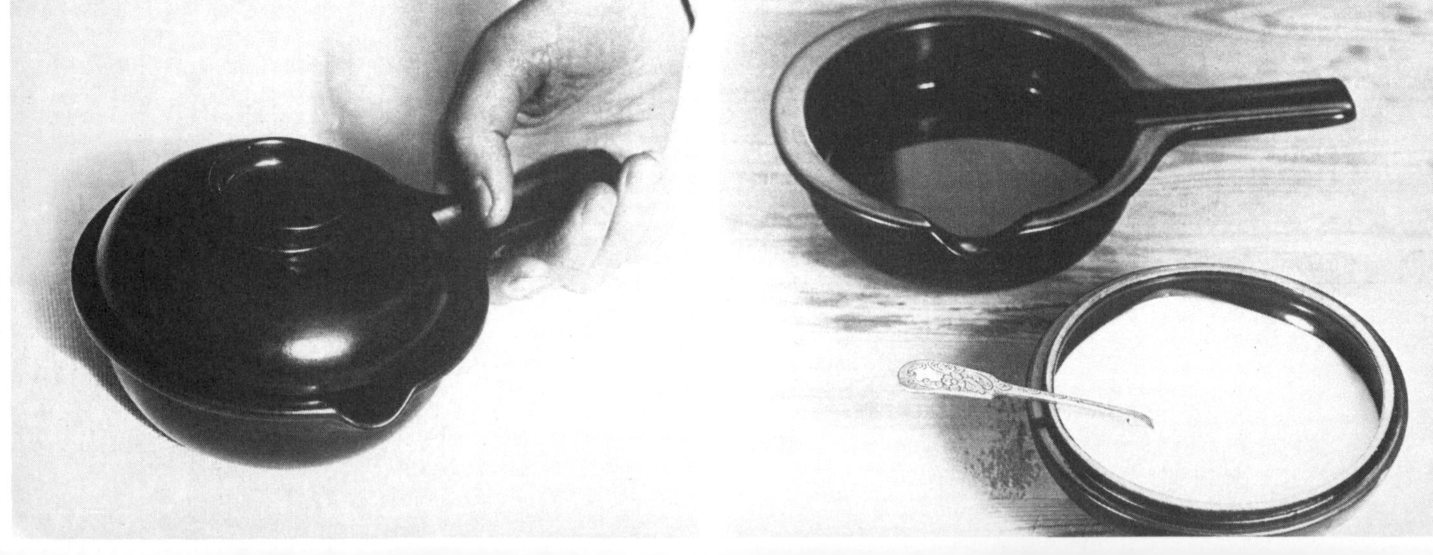

Fireproof, covered cooking pan — part of the "Liekki" series by Arabia. The best thing about this "Liekki" series, and which I believe has made it so popular, is the simple form and the rich, brown color. The pans can be used both on top of the stove and in the oven, and the lid is supposed to have a double function as a sauce or cream dish. Although I agree with many things about this series, I feel it can be greatly improved upon. For one thing, as the photographs illustrate, the handles are so short they are almost useless. The housewife first must risk burning her hand because the handle is not insulated. Next, she must face the possibility of spilling the contents of the pan because she cannot conveniently grip the handles. The lid also needs improvement — it is difficult to hold onto, and not well balanced when it is used as the sauce dish it is "designed" to be. Just how this series has managed to survive production for so many years without someone at Arabia insisting upon these improvements seems a total mystery. As it now stands, and has thus far survived, the form may well be "aesthetic", but it has an aborted function.

Nesting pans by Arabia — also part of the "Liekki" series. These flat pans are also fireproof and intended either for use on top of the stove or in the oven. In many respects, this form is an excellent frying pan — it nests conveniently for storage, and although it is missing a non-stick cooking surface, the pans are quite easy to clean. This series has been in production for many years, but yet the short, non-insulated side handles have never been improved upon. The side handle on the smallest pan is 1 $^{1}/_{2}$ centimeters long; the side handle on the largest pan is only 2 centimeters long. These extremely short handles, when combined with a heat conducting handle material obviously make the pans (particularly the larger sizes) difficult to hold onto when they are hot and filled with food. Another missing feature is that if one wants a cover while frying, none is available in the set. Perhaps the housewife can look forward to an Arabia improvement on this series in the future?

I have included the photographs on these two pages in order to illustrate how a basic form can result in a small industry. Kaija Aarikka began in business as a textile designer — she began by making dresses which also, incidentally, included handmade wooden buttons. This simple round button started a small industry exploring a circle. As the photographs illustrate, Aarikka has not only explored the circle for every concievable household use in wood, but has extended the exploration into metal trays, children's forms, silver jewelry, and finally in plastic balls for room screens.

Stainless steel cooking pans by Hackman. As far as price, these pans seem to be the Cadillac or Chrysler of Finnish cooking utensils. They have all but priced themselves beyond the reach of the average household budget. Only if you can afford to buy the whole series can you then take advantage of the stacking conveniences and the interchangeable covers. When comparing this series with the Danish enamel over iron pans in a similar price bracket, they fall far behind in a number of basic conveniences. For example, it takes much longer to boil a pan of water in the Hackman series (even in the pan with the copper bottom), than in the enamel on iron series. The cooking surface on the stainless pans are also difficult to keep clean — not nearly as convenient as the Arabia "Liekki" series. The stainless is not so stainless. It collects chemical residue, especially from boiling water. Many other foods, particularly with a sauce base tend to stick and burn to the sides of the stainless. The Hackman series is "designed" with pouring lid covers, but these do not lock on — they must be held in place with the hand. If the pan is hot, for example, filled with a vegetable, then when you try to pour off the water through the cover spout, the steam from the water is likely to burn the hand that holds the cover in place.

In testing many of these Finnish cooking utensils, one quickly gets the impression that industry seems to be doing only half of its job. For example, no single industry produces one series that incorporates all the normal convenience features one would expect on a cooking pan. If a customer wants a pouring lid cover, then he must buy a pan from the Hackman series. If he wants a locking lid, then he must buy a pan from someone else that does not pour. If he wants a non-stick cooking surface, then he must buy a third firm's product, and on it goes — with of course the consumer paying the final bill for industrial short-sightedness. It seems incredible that in the 1970's, with at least a dozen Finnish firms producing articles for cooking in many materials — that at least one of these firms could not come up with a cooking pan that met all the convenience features to aid cooking. Hackman, Arabia, and many of the other firms could learn a great deal about their own product design simply by having their cooking pans tested by the non-employees who are supposed to buy them.

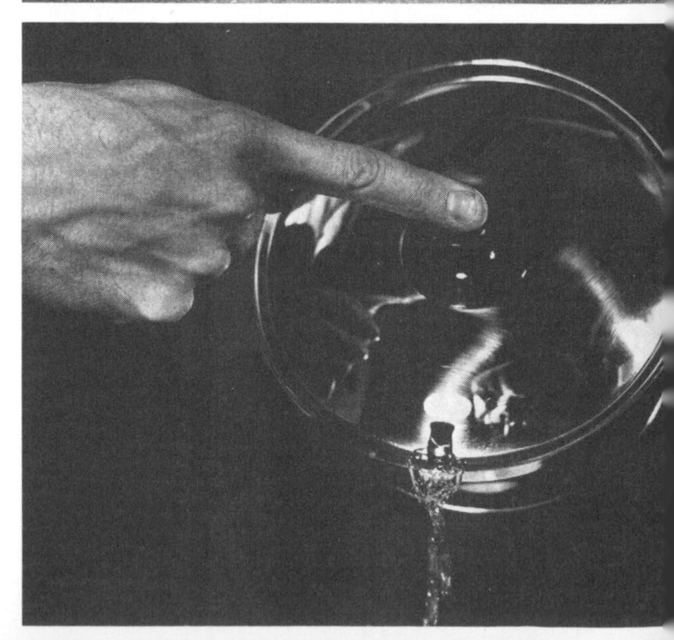

ɔntrary to what most magazines, and
most all of the export publications of
nland would have us believe, every Finn in
nland does not exclusively eat crayfish,
cklig pig, Cornish game-hen, or
arcoaled pheasant. Many hundreds of
ousands of Finnish people eat and enjoy
nple, inexpensive foods such as
lostaja pea soup. And what is wrong
th a nation of pea soup eaters? Perhaps
where else in Europe is pea soup almost
national tradition. And it is a delicious
a soup — perhaps the most honestly
anufactured pea soup in all the world.
ery can, with only an accidental
ception, has ham in it!

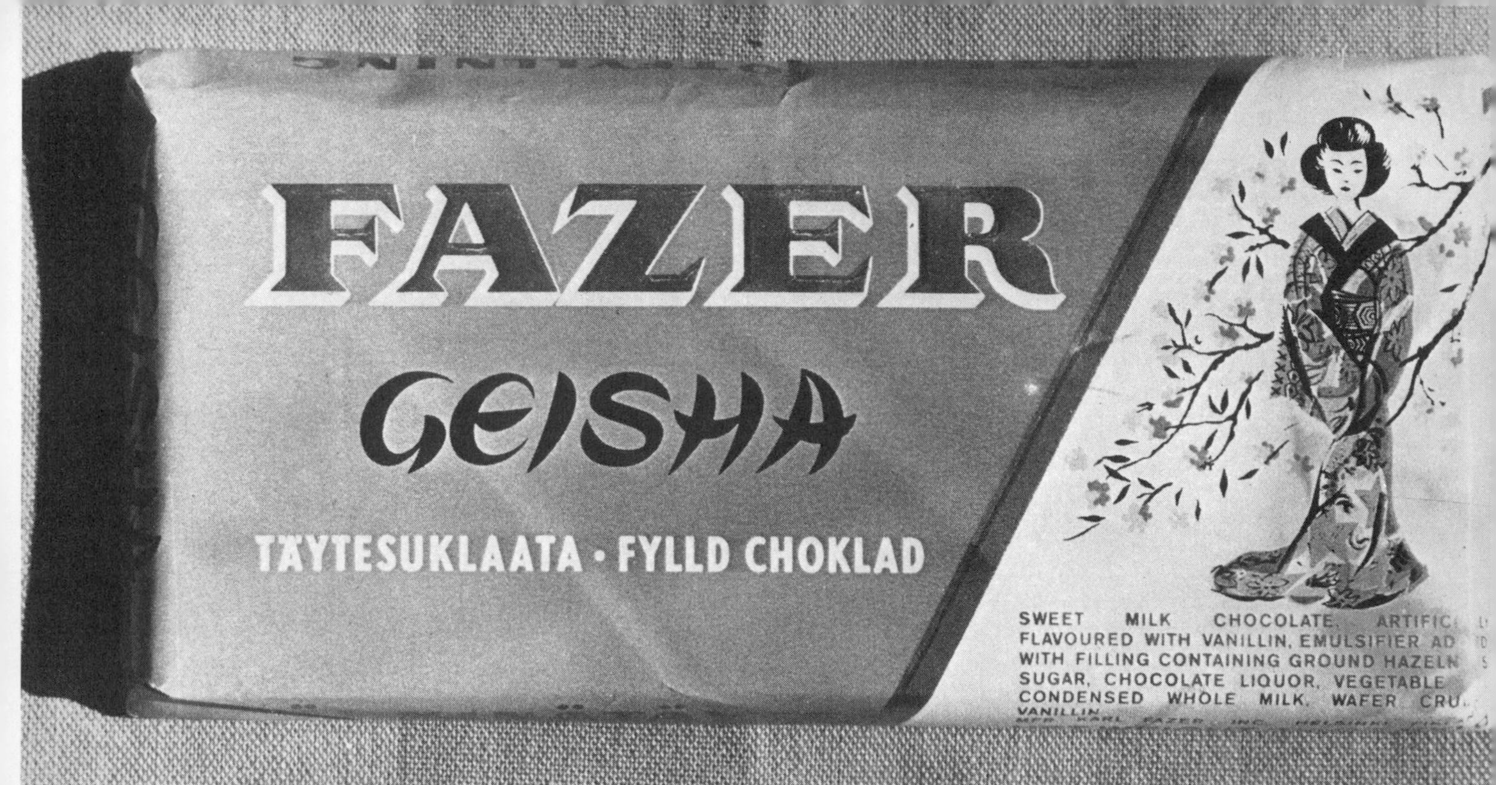

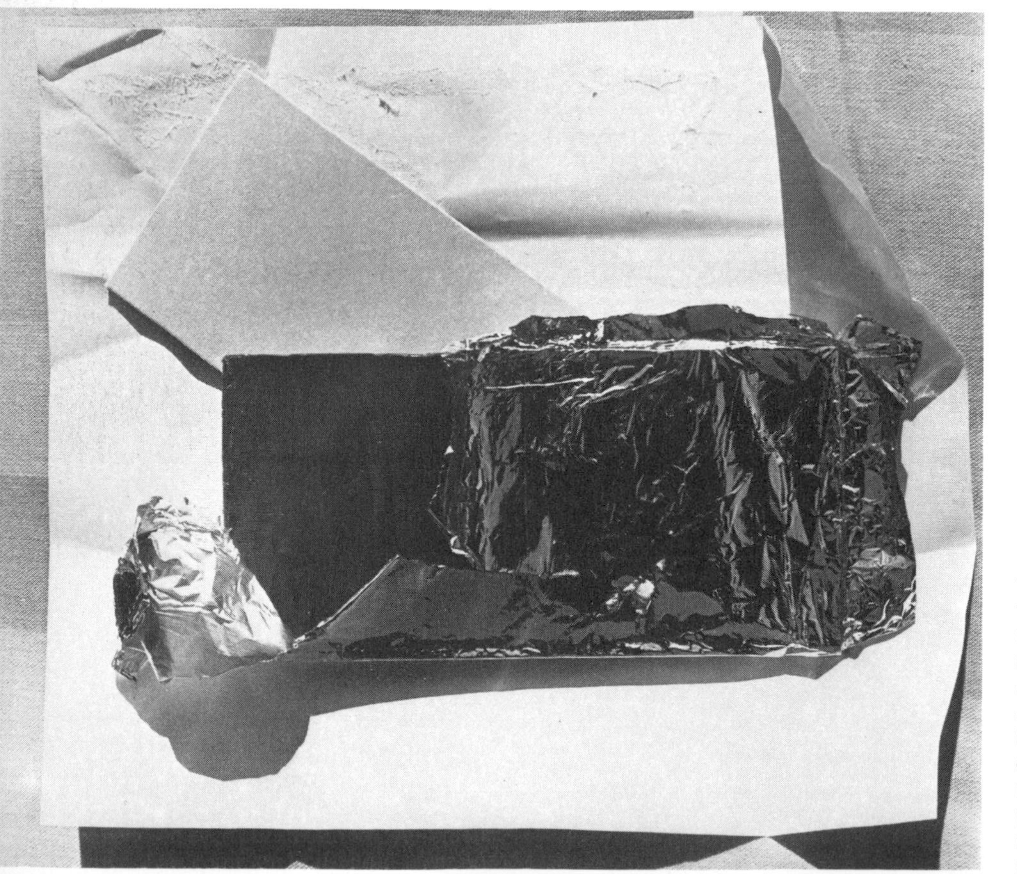

"Geisha" chocolate bar by Fazer. The only thing missing from this chocolate bar to ma it a nearly perfect product is some type of simple opening device similar to that fou on cigarette packages. Otherwise, the chocolate is wrapped in aluminium foil to insure its freshness, and backed with a cardboard insert which keeps the chocolate from breaking. The inclusion of this simple piece of cardboard seems to me what responsible packaging is all about. Fazer h not forgotten its chocolate customers!

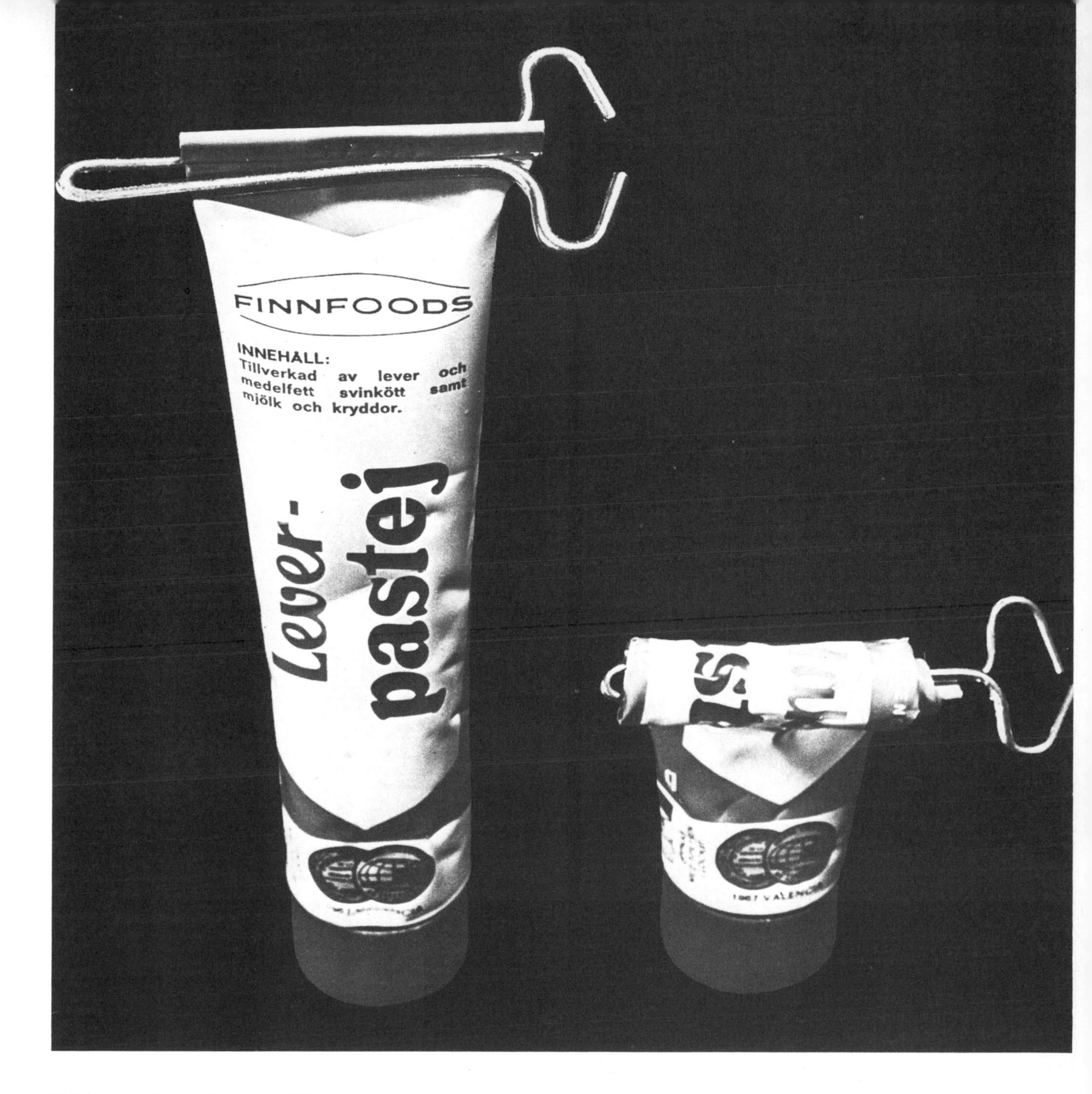

"Maksapasteija — liver-paste" by Finnfoods. The addition of the inexpensive key makes this a near to perfectly packaged product. By twisting the key, the consumer is able to enjoy even the very last drop. Why is it that if Finnfoods has the good judgment to supply a winding key on liver-paste, we cannot find similar conveniences on all of the dozens of other tube products we use in daily life — including toothpaste?

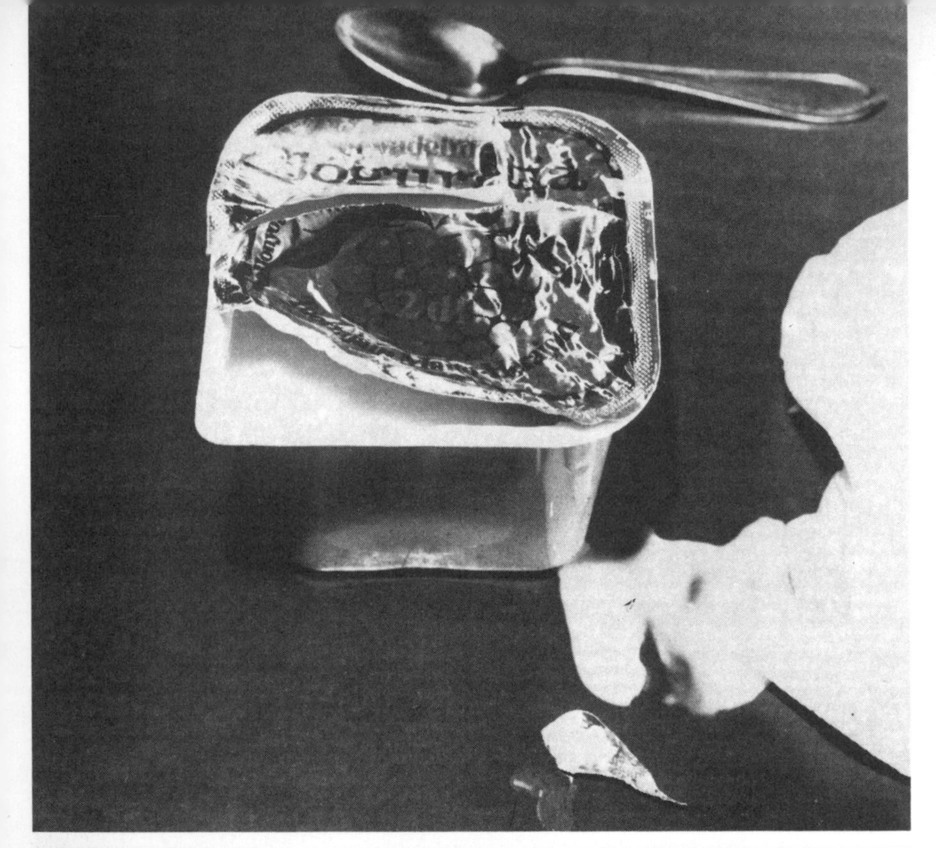

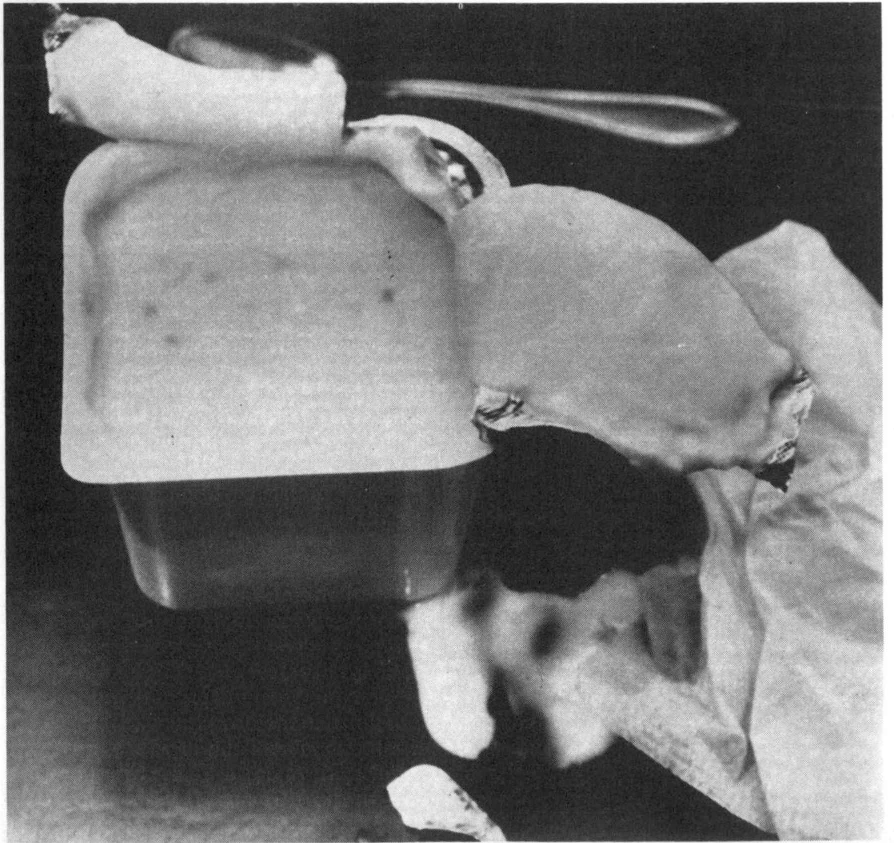

Fantastic! Man has reached the moon, but yet the packaging experts of Finland cannot design a yoghurt package. Every morning, all across Finland, thousands of people confront some of the best tasting, real fruit yoghurt in the world, but packaged within some of the worst containers in the world. It takes the sensitivities of a brain surgeon and the quickness of a magician if one hopes to open a yoghurt cover without slopping yoghurt on both the fingers and table. The foil cover which is supposed to lift up almost always tears into slivers of sticky glop. Can anyone deny room for improvement?

Knoll Associates chair by Eero
Saarinen. Assuming that
nan must always have
functional aesthetic forms to sit
upon — then this form by an
American-Finn leaves very
ittle room for improvement.
In both its base, and its chair,
t has probably become the
nost imitated furniture form of
the twentieth century. There
are hundreds of imitations —
a few even in Finland,
but only one original.

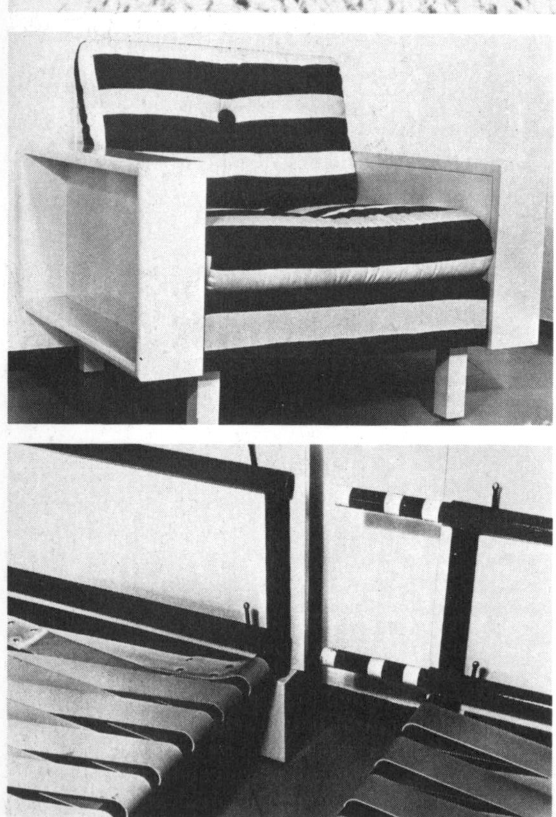

Extendable armchair-sofa by Päivi Harmia and manufactured by Rintala. As the detail photographs indicate, the internal pipe construction of this form makes it extendable by simply removing the two end pieces and inserting an additional center section. This open-ended form can therefore be built into a whole sofa wall, or reduced to a single unit.

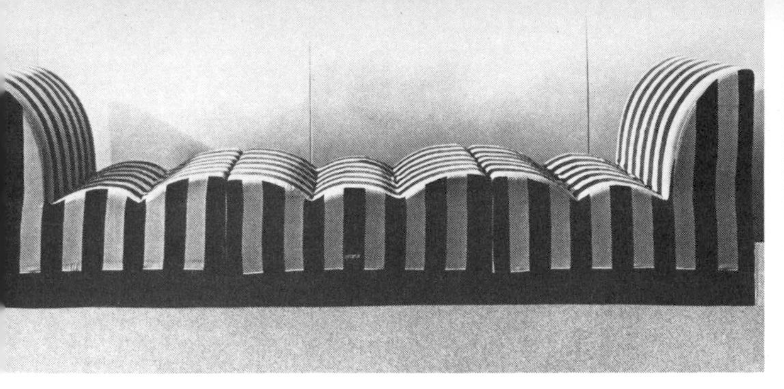

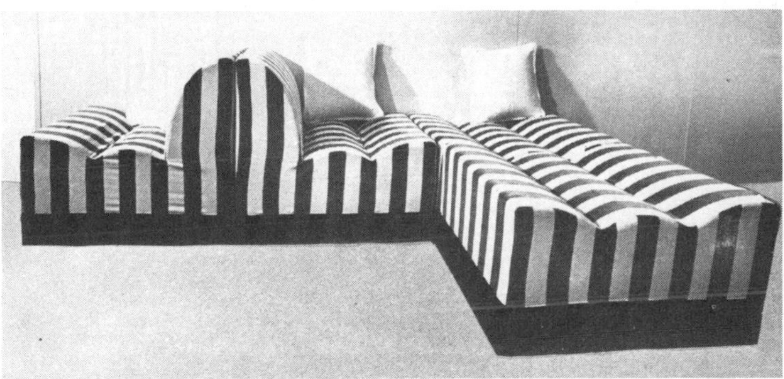

m rubber sectional sofa by Totti Laakso and
ufactured by Oy Skanno. Designer Laakso
eived the form so that it could be deliberately
icated for a budget-conscious, and youthful
ket. As the detail photograph indicates, the
n consists of laminated layers of inexpensive
n rubber which have been cut to form and
red with a zippered textile. This built-in
licity could have made this an inexpensive,
tiple use form, but the producer decided to
re the designer's idealism and market the
n as a luxury status symbol in excess of normal
k-up figures. Mr. Laakso originally
eived the form with a flat sitting surface.
producer added the contours as a means of
ing textile tension in the seat cover. Apparently
kles have no place on a status symbol.

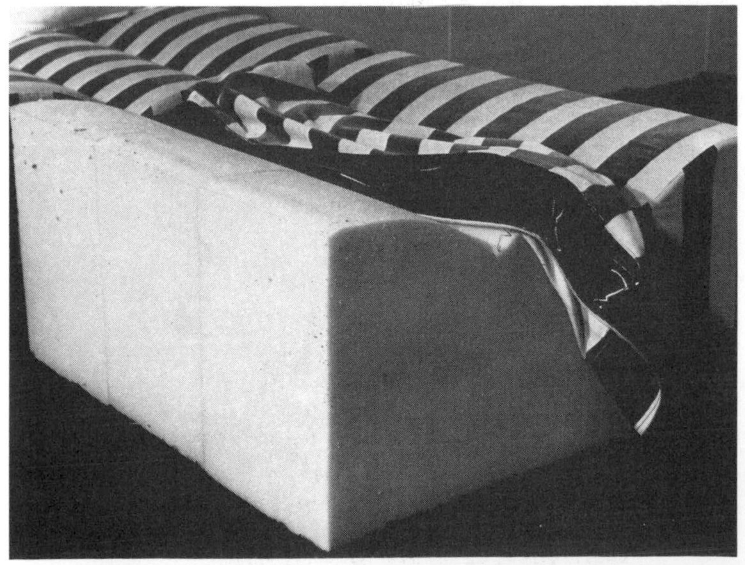

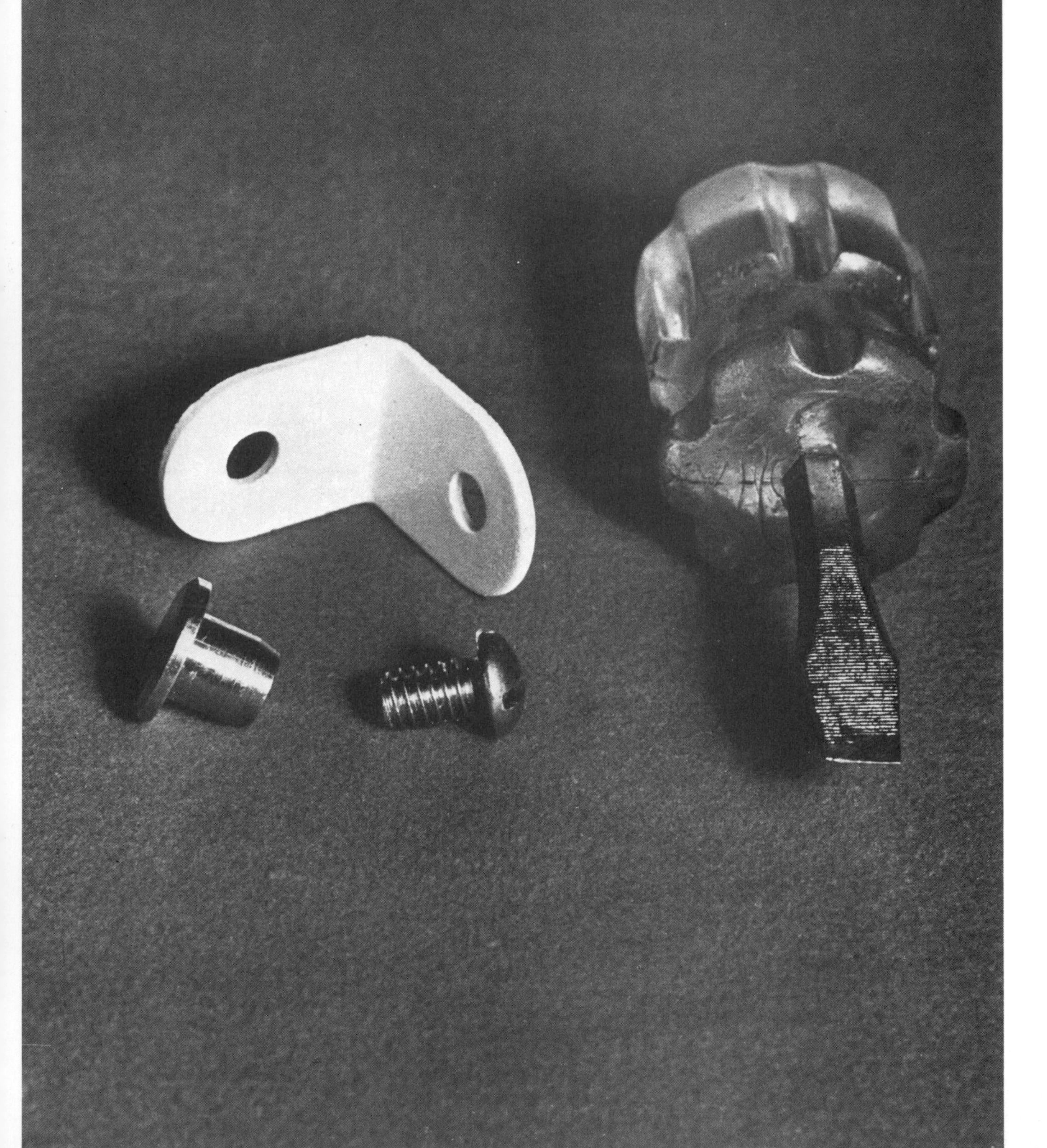

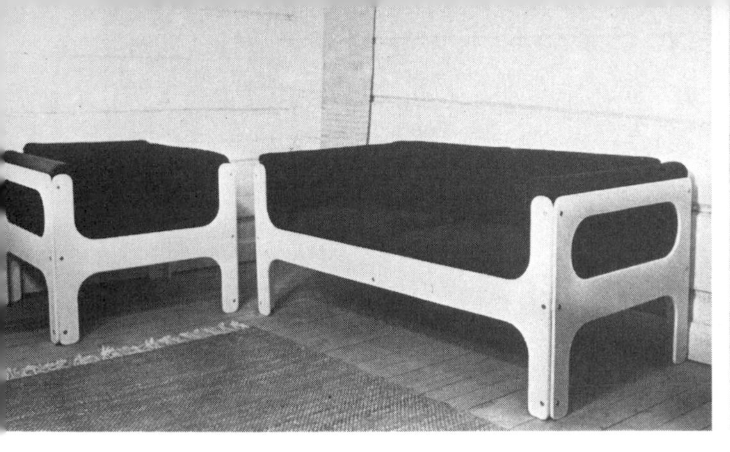

nikka furniture series by Kristian Gullichsen and nufactured by Noormark Handicraft. Simplicity of nstruction, economic use of raw material, and convenient ckaging are the principal features behind these versatile ms. The forms were first manufactured in plywood and n spray-painted with "Dicco-plast" pigment. The nted plywood surface often developed surface cracks. wood has now been replaced with a fiber-board ich does accept paint without cracking. As illustrated, form parts are so well planned that Mr. Gullischsen orts only a 15 % waste of raw material. The construction nciple involves a simple "L" bracket with a brass ew and a brass-capped nut. The table tops in the ies rest loosely on the table bases, and are covered h laminated formica on both sides which makes them ersible.

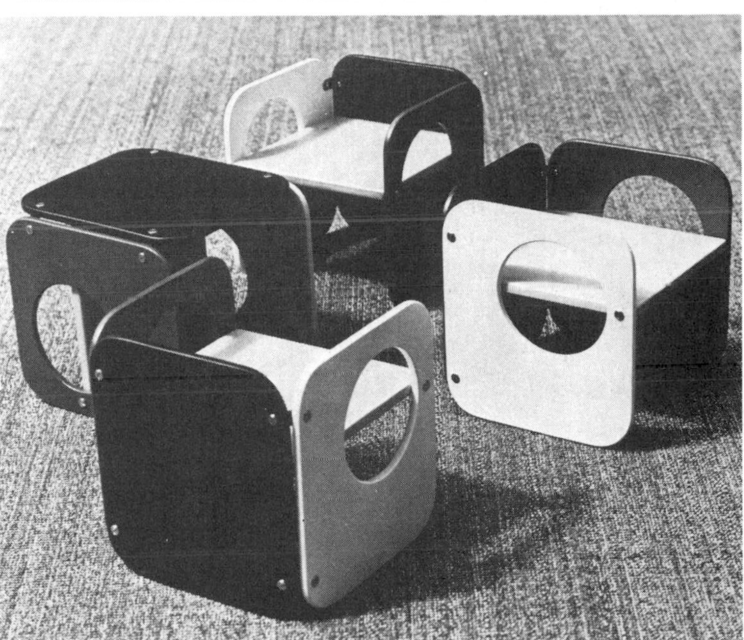

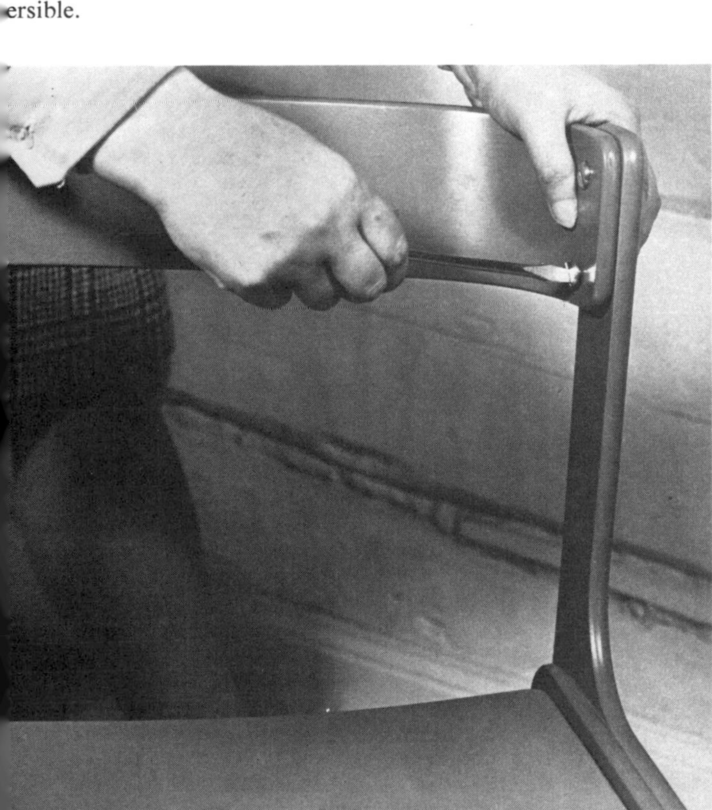

As an alternative to kitchen cupboards, Ristomatti Ratia has used the same net-like framework in his hon that is being used in the Marimekko shops in Helsinki for hanging and displaying garments. This particular ceiling framework is bracketed and screwed into the side walls. It not only allows one to lower an otherwise high ceiling, but as the photographs indicat this means of storage also allows for maximum flexibility in room arrangement, plus instant access to the many objects one accumulates for use in a kitchen, but often forgets about. This net-like framewo in the Ratia home is so strong that it supports a child's swing. The construction concept of the framework will also allow one to hang portable cupboards for the home-owner who wants to hide wh he has, or prefers to protect his inventory from dust. If the idea were industrially developed along the lines of the expanding pole lamp with the built-in tension, the idea could be used by average home-owners who felt a need for an alternative to kitchen cupboards.

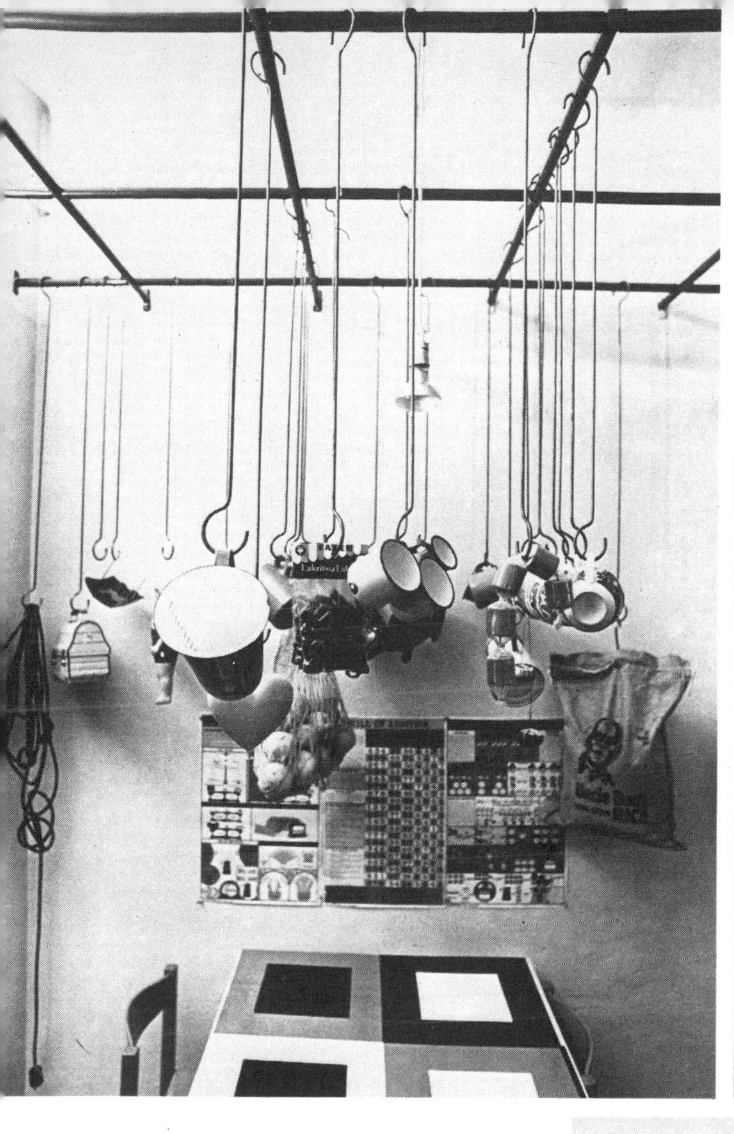

JAFFA ORANGES

JMC
67

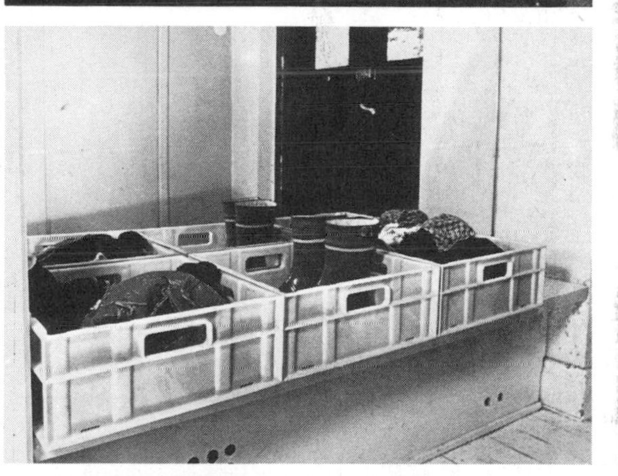

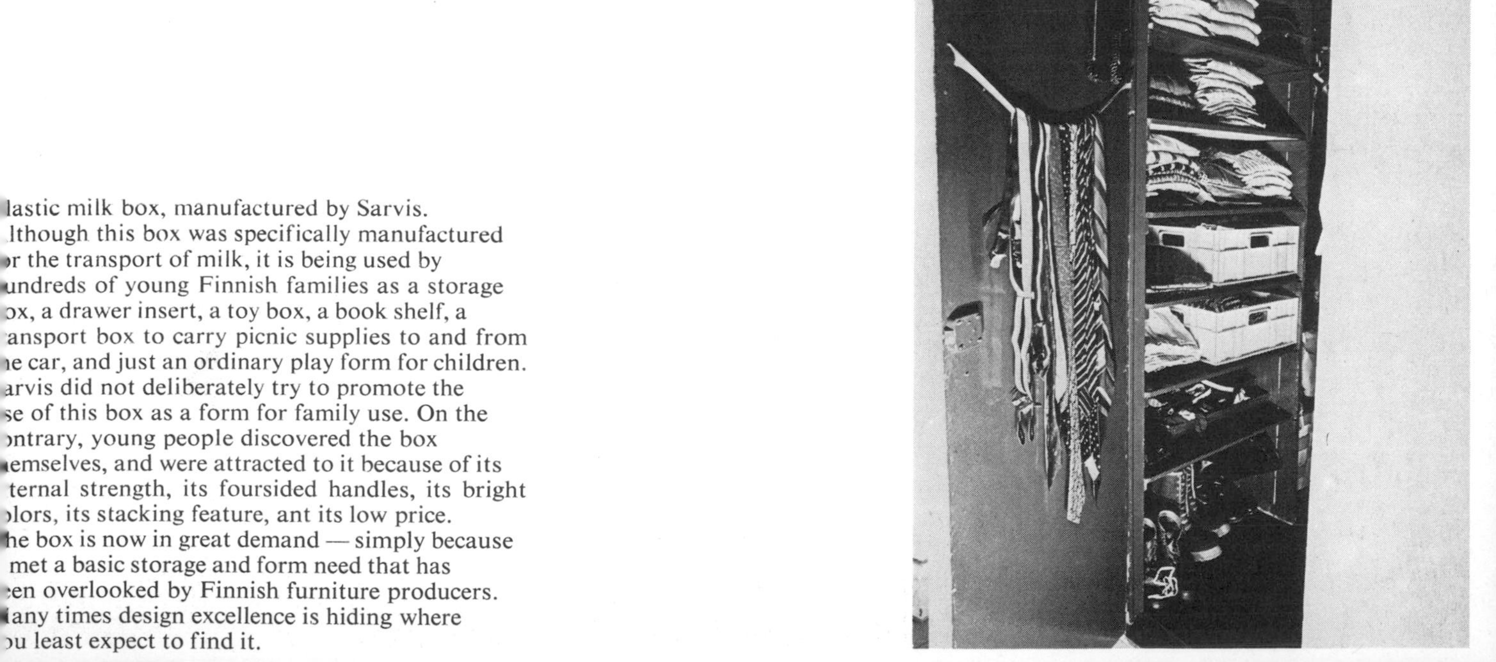

lastic milk box, manufactured by Sarvis.
lthough this box was specifically manufactured
or the transport of milk, it is being used by
undreds of young Finnish families as a storage
ox, a drawer insert, a toy box, a book shelf, a
ansport box to carry picnic supplies to and from
ne car, and just an ordinary play form for children.
arvis did not deliberately try to promote the
se of this box as a form for family use. On the
ontrary, young people discovered the box
emselves, and were attracted to it because of its
ternal strength, its foursided handles, its bright
olors, its stacking feature, ant its low price.
he box is now in great demand — simply because
met a basic storage and form need that has
een overlooked by Finnish furniture producers.
any times design excellence is hiding where
ou least expect to find it.

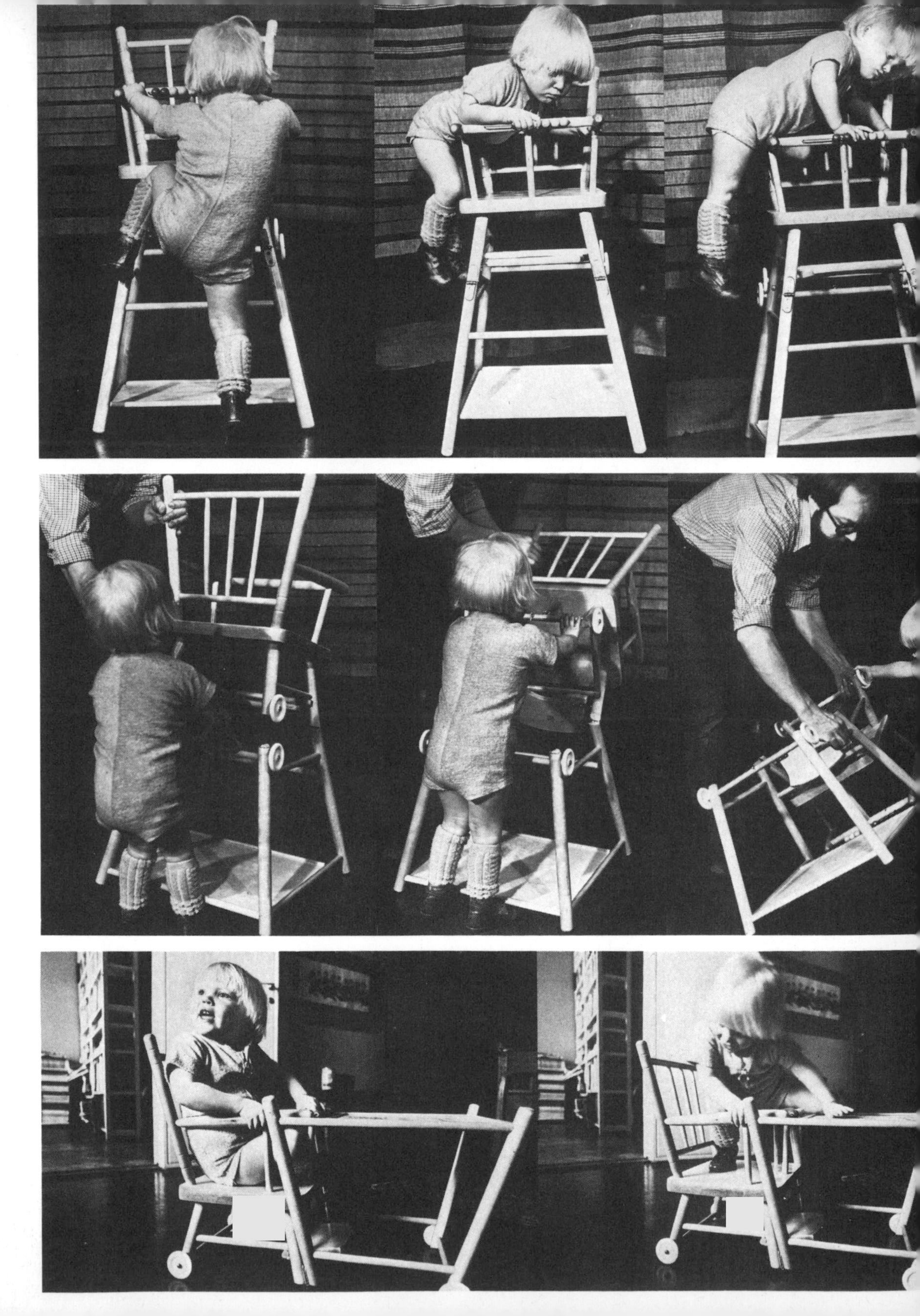

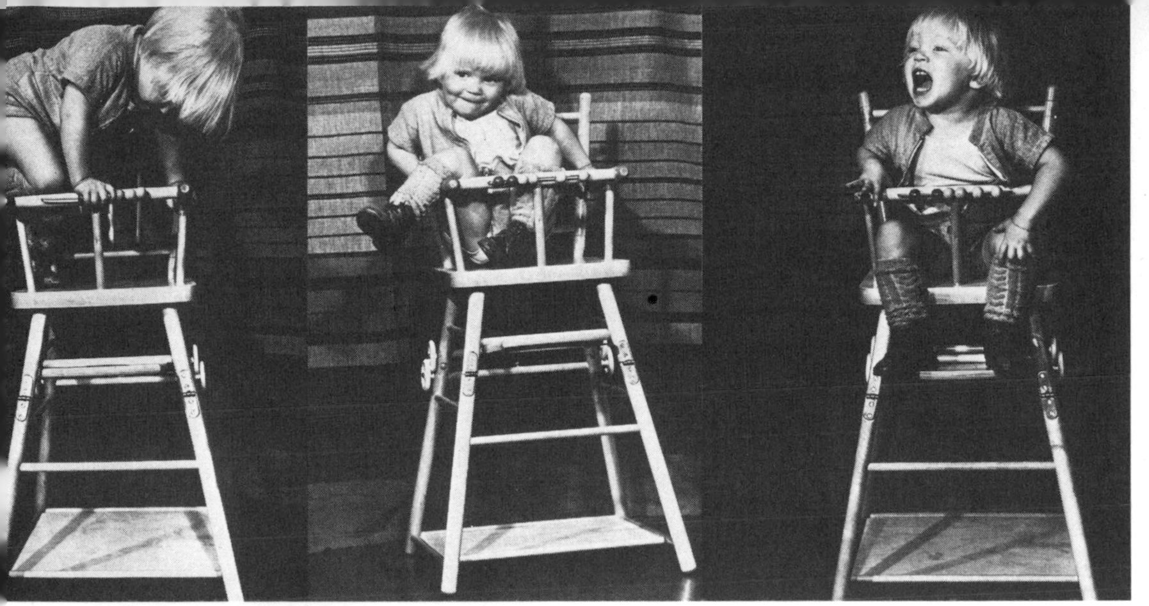

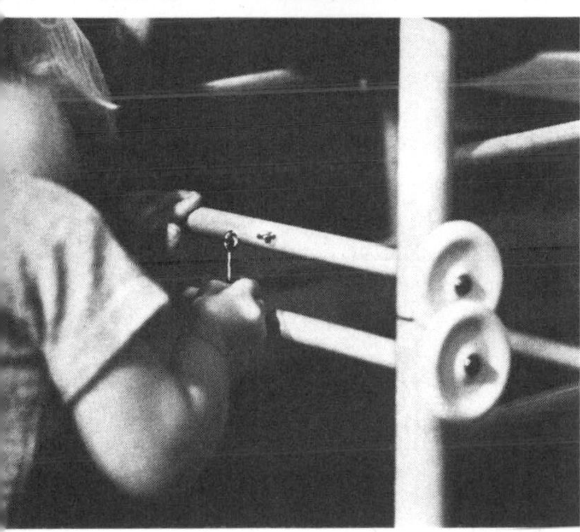

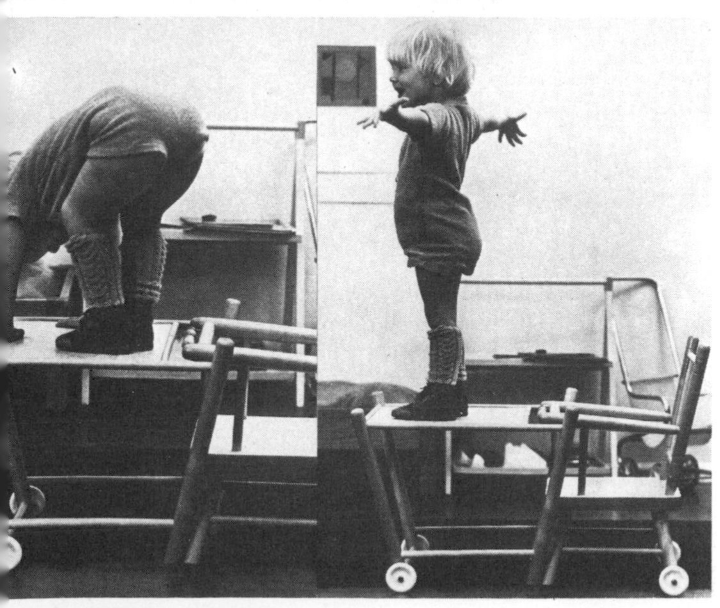

This almost classic Finnish highchair, manufactured by Niemen Tehtaat Oy, is another of the objects which have seldom been illustrated among forms publicized as "Finnish design". The form is basically the same as it was twenty years ago. As a form, it is durable, it is strong, it is inexpensive, it is extremely well balanced, it has a minimum of sharp edges, it quickly and conveniently converts into a low chair and table, and it has its own set of wheels under it. The chair may not be "aesthetic" according to sophisticated modern standards, but it nevertheless has proved itself as both rational and honest. Mila had never seen the chair before I brought it into her home. She immediately, on her first try, was able to completely conquer it — even to the locking of the hook. It is interesting to note that most modern highchairs are manufactured with an eating tray which is attached to the form. This Niemen Tehtaat highchair has no such tray. It is encouraging to see that this firm still considers children as a part of the family — and has refused to segregate them into their own self-contained eating environment.

Bunk bed by Pirkko Stenroos, manufactured by Muurame. This wood-frame bunk bed is an open-ended form which provides multi-functional advantages to Finnish families with children, or with summer houses, or even with problems of limited living space. The bed units stack one upon another. An optional hanging bookcase can be added to the end of the bed, and optional clothing hooks can also be hung along the sides of the be frame. Mrs. Stenroos designed a series of forms which are based upon this same open-ended concept. The drawers slide out from under the bed on wheels and can quickly be converted into a toybox car equipped even with steering wheel. It is no surprise that since this form rationally answers the demands of a growin family, it has become one of the most popular furniture forms in Finland.

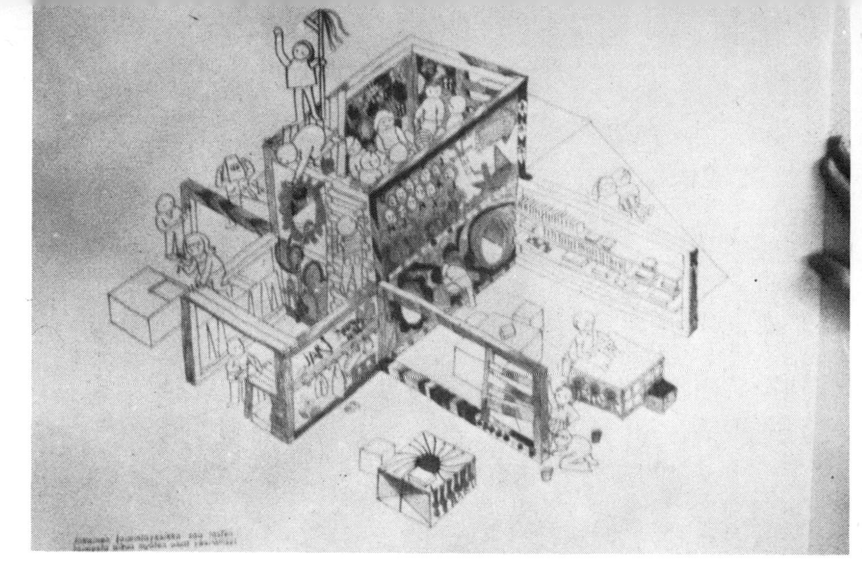

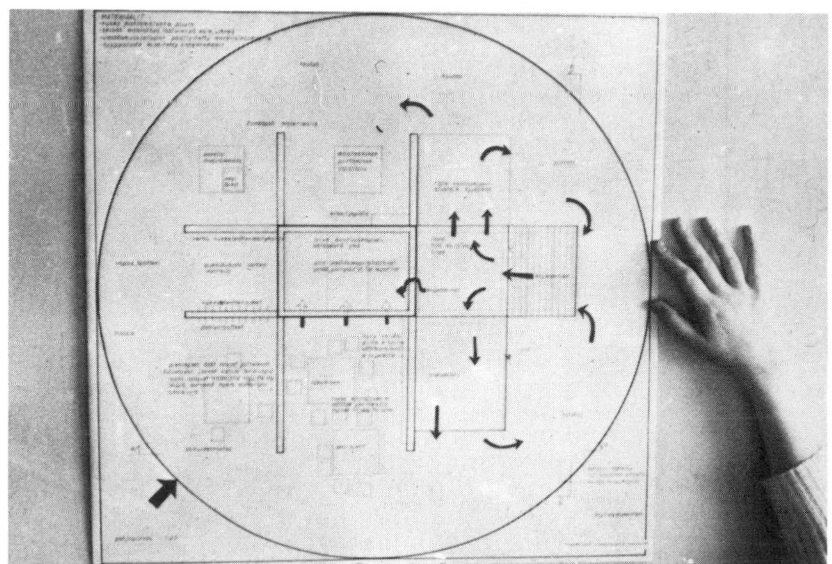

rototype for activity environment, designed
y Anna Tauriala. The unit is intended for
dustrial production and for use by
eighborhoods, or one single unit can even be
dded onto for larger groups. The concept is
framework for children from 3 to 8 years old.
he environment consists of multiple play
reas using multiple materials — for example,
usic, reading, theatre, art, and areas for
ich physical activites as climbing, sliding,
iding, and even resting. Mrs. Tauriala feels this
nit is an alternative to institutionalized
laygrounds and she has tried to deliberately
llow for the unexpected free expression of
hildren, as opposed to grouped and pre-
rogrammed activity. The basic form is also
ortable and can be set up both indoors and
utdoors. As of this writing, and even
ough this prototype fills a desperate activity
eed — the prototype has gone by almost
nnoticed.

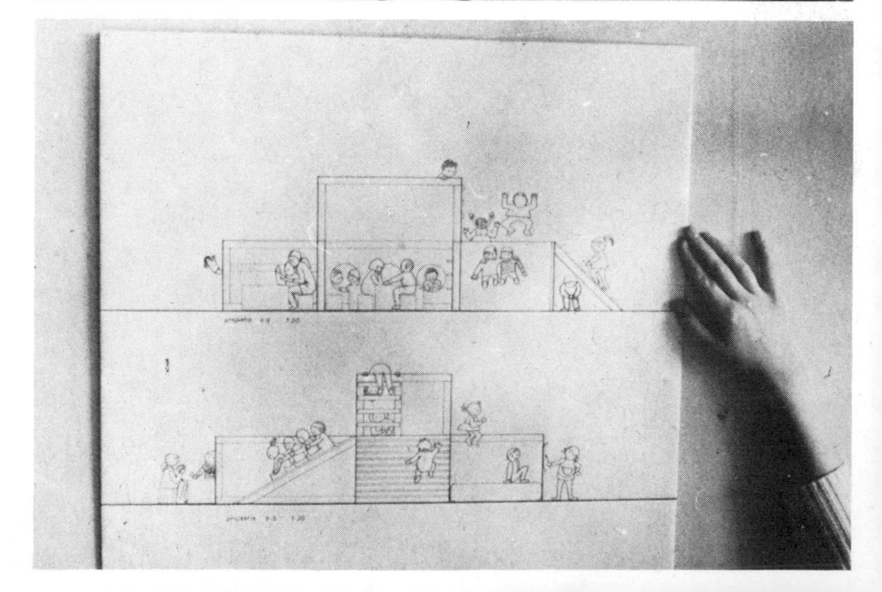

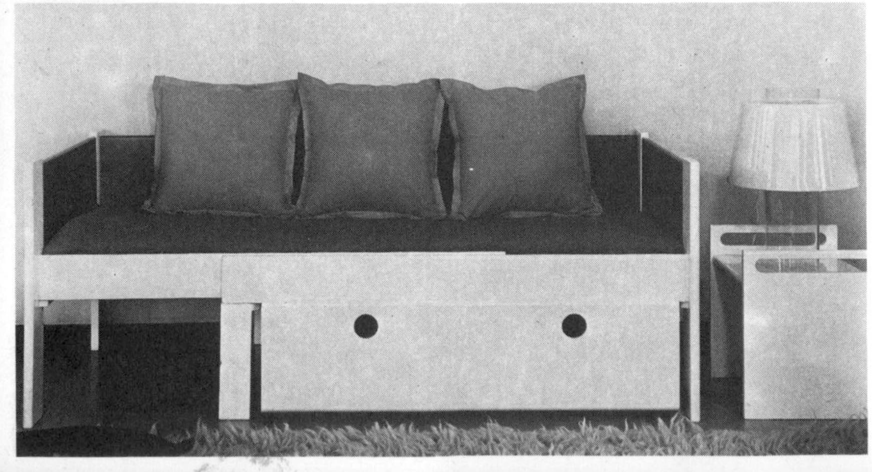

Multiple use child's bed by Päivi Harmia, and manufactured by Rintala. Mrs. Harmia has been deeply concerned with multiple-use forms for budget conscious families. The form was concieved so that it could be used as a high-sided baby bed, extended to a low-sided junior bed, and then extended again into a full-sized sofa. The wooden sides are interchangeable, and the form was concieved to use either two storage drawers when used as a child's bed, or one large storage drawer when used as a normal sofa. The small drawers are supposed to be made with optional wheels and an interchangeable drawer cover that converts the drawer into a rolling car.

This form was so well received by a budget conscious public that the small firm of Rintala was literally swamped with orders. This rush of business, and the economic incapacity of the producer to expand his facilities has resulted in confusion surrounding the multiple-use concept. At the time these photographs were made in the Rintala showroom, the form was being displayed and sold as an extendable junior bed, and the high sides were not even available. Just in order to record Mr. Harmia's full concept, we had to wait over 6 weeks to get just one complete unit. Although the drawers were supposed to have pivoting wheels under them — they were being produced with only stationary wheels. The demand for the form had put such a pressure on the producer that in the end, the customer was only exposed to part of the total concept.
As a result of this Rintala experience, and several other similar experiences with multiple-use forms — it appears obvious that the demand for rational furniture forms in Finland is many times greater than industry even suspects.
Even despite this demand, most firms continue to exend themselves even further into "aesthetic" luxury furniture — and seem to totally ignore the young, budget conscious buyer. It is indeed encouraging to note that firms like Rintala, Muurame, Nuuponen, and several others have given their support to designers with a working social conscience.

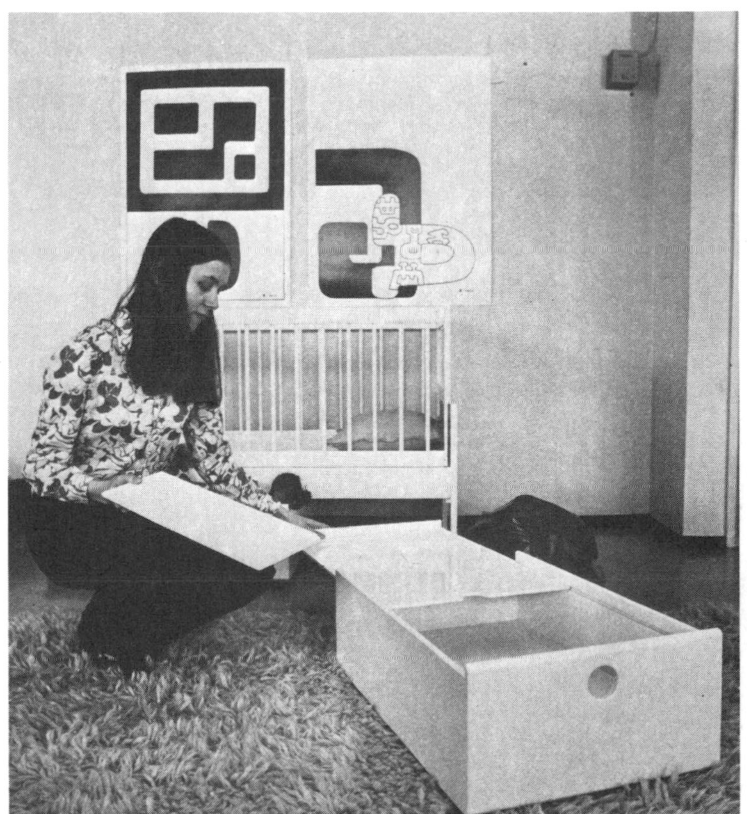

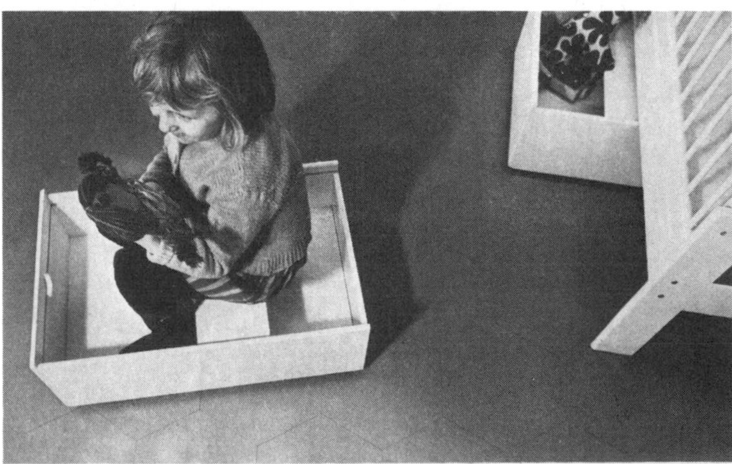

"Poppa" double-bench form by Anna Tauriala — and occasionally manufactured by Korhonen Oy. Mrs Tauriala conceived this form so that it could be made out of any inexpensive strong material including plastic. The form was deliberately conceived with an open-ended, multiple-use function as either an adult furniture form, a form for a children's room, or a form for use in schools. As these drawings by Mrs. Tauriala indicate, the form does have endless possibilities — from using it as a table base — to using it as a sickbed stand — to using only half of the form as a scooter or sled. I firmly believe this is one of the strongest, but yet underrated forms in Finland, and one of the clearest open-ended forms to come alc in many years. Its production in Finland has been plagued with difficult even though the form has received positive critical support on all sides, a is in concept, a very simple form to fabricate. At the present time, Korhonen Oy has been ocassionally manufacturing the form and retailing it

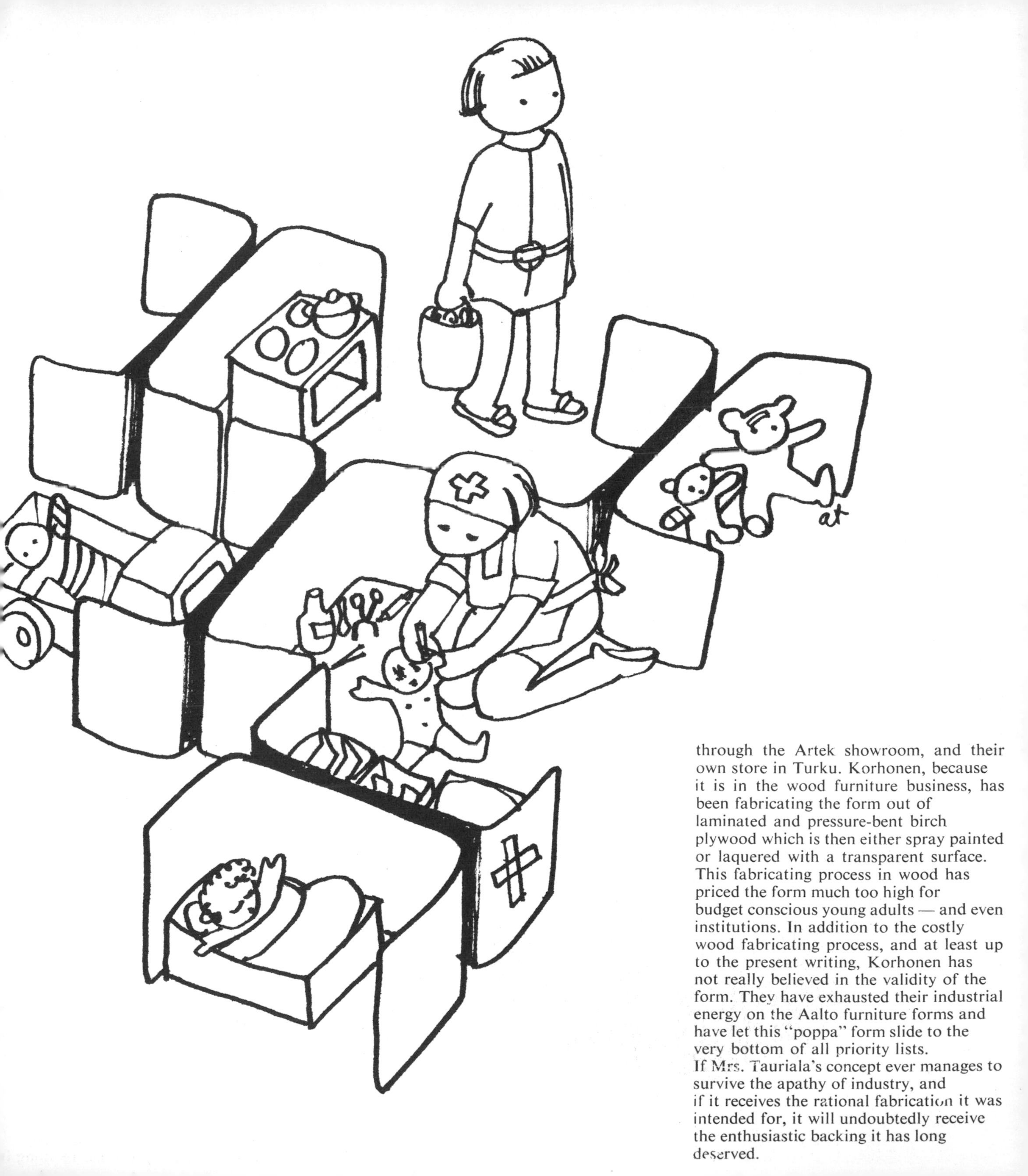

through the Artek showroom, and their own store in Turku. Korhonen, because it is in the wood furniture business, has been fabricating the form out of laminated and pressure-bent birch plywood which is then either spray painted or laquered with a transparent surface. This fabricating process in wood has priced the form much too high for budget conscious young adults — and even institutions. In addition to the costly wood fabricating process, and at least up to the present writing, Korhonen has not really believed in the validity of the form. They have exhausted their industrial energy on the Aalto furniture forms and have let this "poppa" form slide to the very bottom of all priority lists.
If Mrs. Tauriala's concept ever manages to survive the apathy of industry, and if it receives the rational fabrication it was intended for, it will undoubtedly receive the enthusiastic backing it has long deserved.

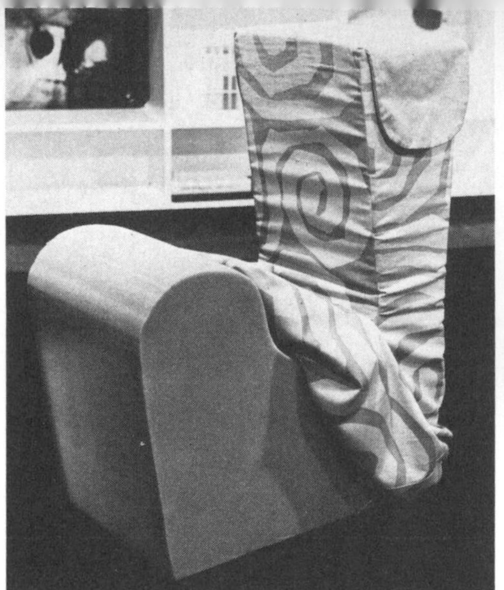

Foam-rubber children's animal form manufactured by Sokos. As the photographs indicate, these forms are sawed from a rectangular block of foam-rubber and then covered with a zippered textile sack. The form is inexpensive, but it also leaves a great deal to be desired as far as the quality of the textile, the care exercised in sewing the textile, and the general balance of the form. Niko, who tried the form out in these photographs, kept tipping backwards. The whole form has a tendency to lean, as if the horse wanted to sit up on its hind legs. It is, however, encouraging to see Sokos explore inexpensive forms for children, and to see their basic good use of a durable and practical concept.

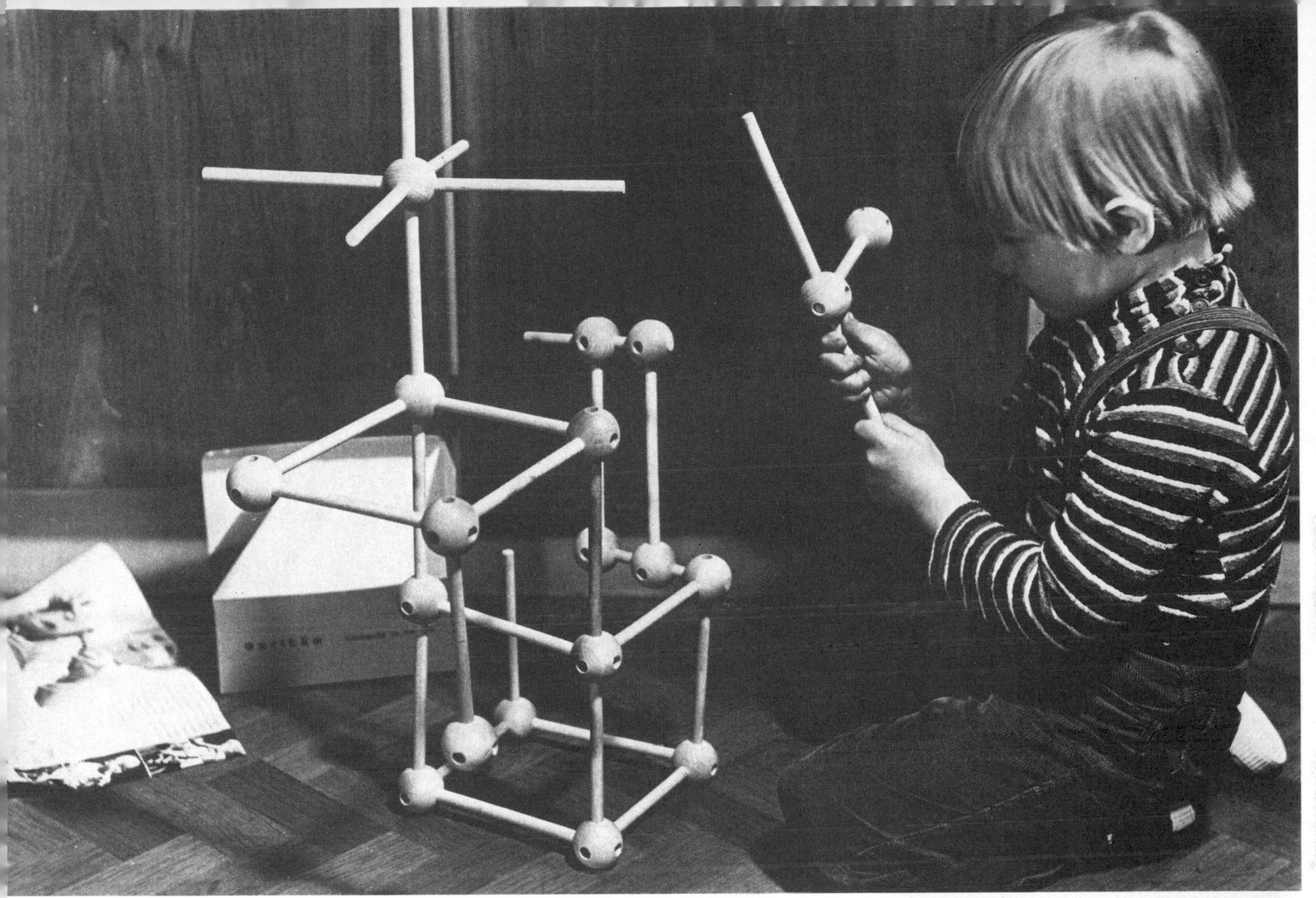

Building sticks by Aarikka. This building toy is represented as Finnish "design", although it is based upon an almost classic idea, with very little variation from, for example, the American "Tinker Toy" which has been on the market for well over 40 years. The basic building principle is a good one, and seems to offer maximum possibilities for the expression of individual form. Five year old Nipa, who tried the toy out while we looked on, encountered several problems with this Aarikka version. The basic problem seems to be that too few pieces are included in the package. This places serious limitations on what a child can do with it. The parent can of course solve this producer problem by buying one or two more sets. This would undoubtedly make Aarikka very happy, but not the parent. Another problem, and a serious one for both the child and the parent is that the toy is poorly packaged. It arrives in a flimsy cardboard box in which it is also supposed to be stored, whereas, for example, the competitive American "Tinker Toy" is packaged in a sturdy, durable cardboard tube with metal, screw-on covers.

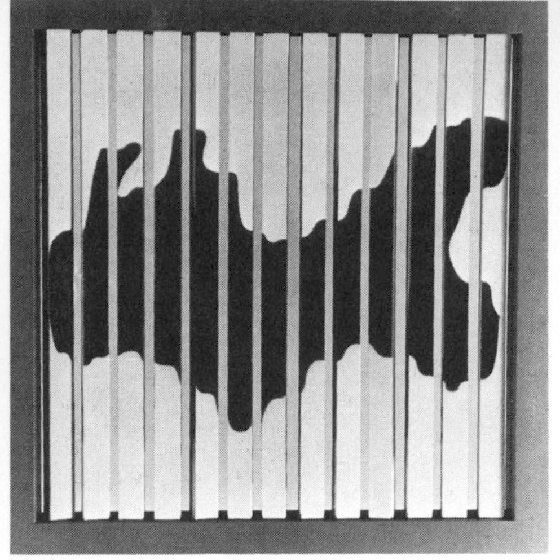

These photographs illustrate optical block toys designed by Helsinki Ritavuori, and manufactured in wood by Oy Kinetik Limited. These forms represent themselves to be "educational" in the sense that they deal with the concepts of movement, or "kinetics". The 1971 issue of "Designed in Finland" magazine tells us that: "toy expert, kindergarten teachers, architects, psychologists, and parents alike have acknowledged the educational, therapeutic, and non-preprogrammed value of kinetics". The same magazine tells us that Mr. Ritavuori's toys "enormously appeal to the imagination and planning sense of children, who find endless do-it-yourself possibilities in moving, manipulating and painting these toys". If we are to believe what appears to be a non-advertised, and responsibly written article in "Designed In Finland", then why is it that Mr. Ritavuori himself told me that these forms were difficult to sell? Mr. Ritavuori explained that because the forms do not resemble reproduced, miniature traditional forms, the parent does not immediately understand the concept behind the forms. Mr. Ritavuori reported that in order to sell the forms, the salesman must be able to explain the concept to the buyer. In other words, the toys must be explained before they can be appreciated. Now if the parent-consumer must be taught why it is intellectually important to buy this form, and if the child has intelligence at least equal to that of the parent, then it must follow that before the child can appreciate the forms, he also must be "educated". Herein lies a basic danger in designing forms for children, especially so-called "educational" toys. To teach a child how to use a toy according to the designer's intellectual concept even before the child can "educate" himself with the toy is not at all the same as allowing the child the freedom to simply use the toy in the way he wants to use it. In addition to the adult intellectualization of this toy (which I'm quite certain the average child is not interested in), the toy is manufactured in wood with the many individual parts having to be painted with as many as four separate colors. The use of wood, and the difficulty of fabricating multi-colored paint, makes this "educational" toy expensive. The forms are designed so that composition parts are both removeable and interchangeable — the child can stack them in several patterns, or glue on photographs as illustrated in two of the factory examples. "Designed In Finland" magazine also reported that these forms: "offer fun and relaxation to grown-ups with an itch to devise and decorate". To this end, Mr. Ritavuori has designed the forms so that they can be hung on a wall strictly as a decoration, or they can be placed within a framework composition and used as a room-divider screen.

This toy then has "education" in it for the child, and "fun and relaxation" in it for the adult. Several children, I assume of average intelligence, tried this form out at our request. If "Designed In Finland" is correct, then it is assumed that the children who tried these forms out must have been problematic — because it could certainly not be a problematic form could it? Have you ever tried to explain the educational, therapeutic, and non pre-programmed artistic value of kinetics to a 5, 6, or 7 year old child? But the toy is a beginning for Mr. Ritavuori. Although the toy is not supposed to be limiting, it is in fact limiting precisely because prescribed patterns are already painted on all surfaces. The forms could very definitely be made non-limiting by simply allowing the child to create his own patterns. This could be done with surfaces either like used on felt pictures, schoolroom chalkboard, or even carbon-erasable surfaces.

Motor movement toys in wood, designed by Jorma Vennola, Pekka Korpijaakko, and manufactured by Aarikka. By adult "aesthetic" standards, these four separate forms have been hailed as the ultimate in "design" excellence, and as a matter of fact, they have recently been awarded a Danish toy prize (the selection was of course juried exclusively by adults, rather than children). These toys represent themselves to be "educational" in that they are supposed to teach a very young child the basic motor movements of threading, twisting, turning, and pulling. As adults, we all apparently agree that these

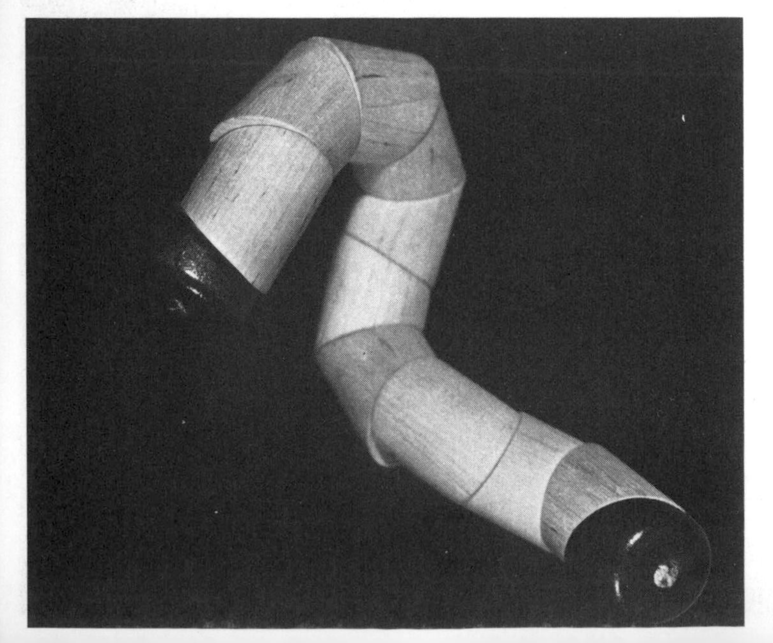

forms are both "aesthetic" and excellent "design", but in passing our judgment, we have somehow overlooked the child. We have become so super-sophisticated that as manufacturers and toy designers, we now inform parents that they must purchase specialized toys which "educate" the child to twist. As conscientious parents, we buy this toy in order not to subvert "education". We do not for one moment realize that we have been victimized by advertising — and that we can in fact give our child a discarded plastic shampoo container (which incidentally costs nothing), supply him with the screw-on cover, and let him learn this exact same motor movement at no expense to anyone. No, twisting has now become a specialized "educational" concept which of course requires a specialized toy. And what about threading? Well, we give our child this "designed" threading toy — a simple pin, a length of cord, and a block with holes in it. And what does the child do with it? Oh, the child? You mean we remembered the child? Well, he assumes that our "educational" threading toy is some kind of a wagon, and so he drops it on the floor and begins to pull it around the room using the pin and cord as a handle. We rush up to him and say: "No, no little one, you don't understand . . . this is an 'educational' toy". So we end up having to "educate" the child how to use the toy before the child can "educate" itself how to thread. And what does the child do once he has learned this dramatic exercise of putting a pin through a hole? Oh the child? Well, he puts the toy aside and finds himself a cardboard box, one of his mother's cooking pans, or something where he can use his own imagination despite his well-intentioned parents. We adults find ourselves in a peculiar conflict with the natural self-interests of our own children. Our children seek to imitate and join our adult world, but we try desperately to keep them out. We give our child a plastic scissors — a miniature reproduction of our own adult scissors, but when the child learns that he has been discriminated against, and that the miniature scissors does not cut paper as well as

ie scissors his mother uses, he prefers the real thing as opposed
ɔ a poor imitation. We give our child a wooden block with
'heels on it and tell him that it is a truck. He looks out the
ʼindow, sees a truck, and knows perfectly well that the wooden
lock with wheels under it is not a truck. The child prefers an
:nglish "Matchbox" miniature scale reproduction of a truck,
ather than our especially "designed" wooden block with wheels.
Ve give our child a cuddly stuffed doll that doesn't look human,
ut makes us quite happy, and we find that the child prefers a
Matel Barbie" doll which does look real. As parents and adults,
ve seem to place most our energy in trying to change the natural
ıstincts of the child's world, rather than in trying to improve upon
ur own world. We want to save the child and insulate it from our
ften cruel adult world, but our energies are in direct conflict with
he interest of the child. The task of designing toys is not easy —
t requires finding out what children want rather than what adults
vant. It requires a careful protection of childhood imagination,
ather than exploiting it by imposing an adult "aesthetic". It is
mportant to note that there are dozens upon dozens of common,
rdinary household objects which are more fascinating to
. child than many of the so-called "specialized educational"
oys we have deliberately managed to "design". These Aarikka
orms may well meet adult standards of what toys should look
ike and how they should "educate", but they fall short when
ne considers a child's world through a child's eyes.

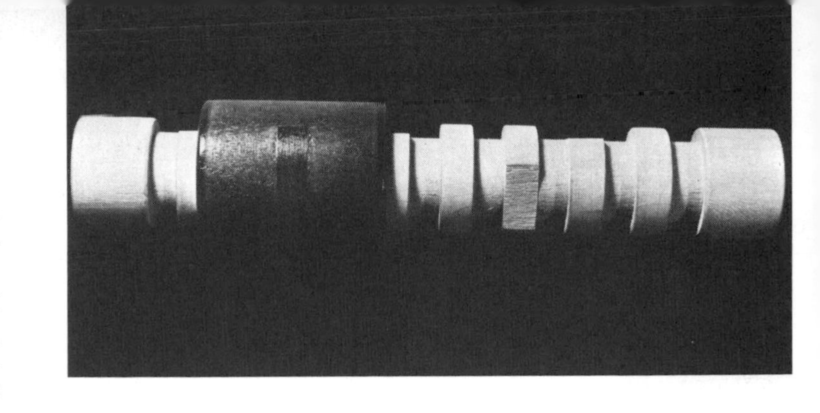

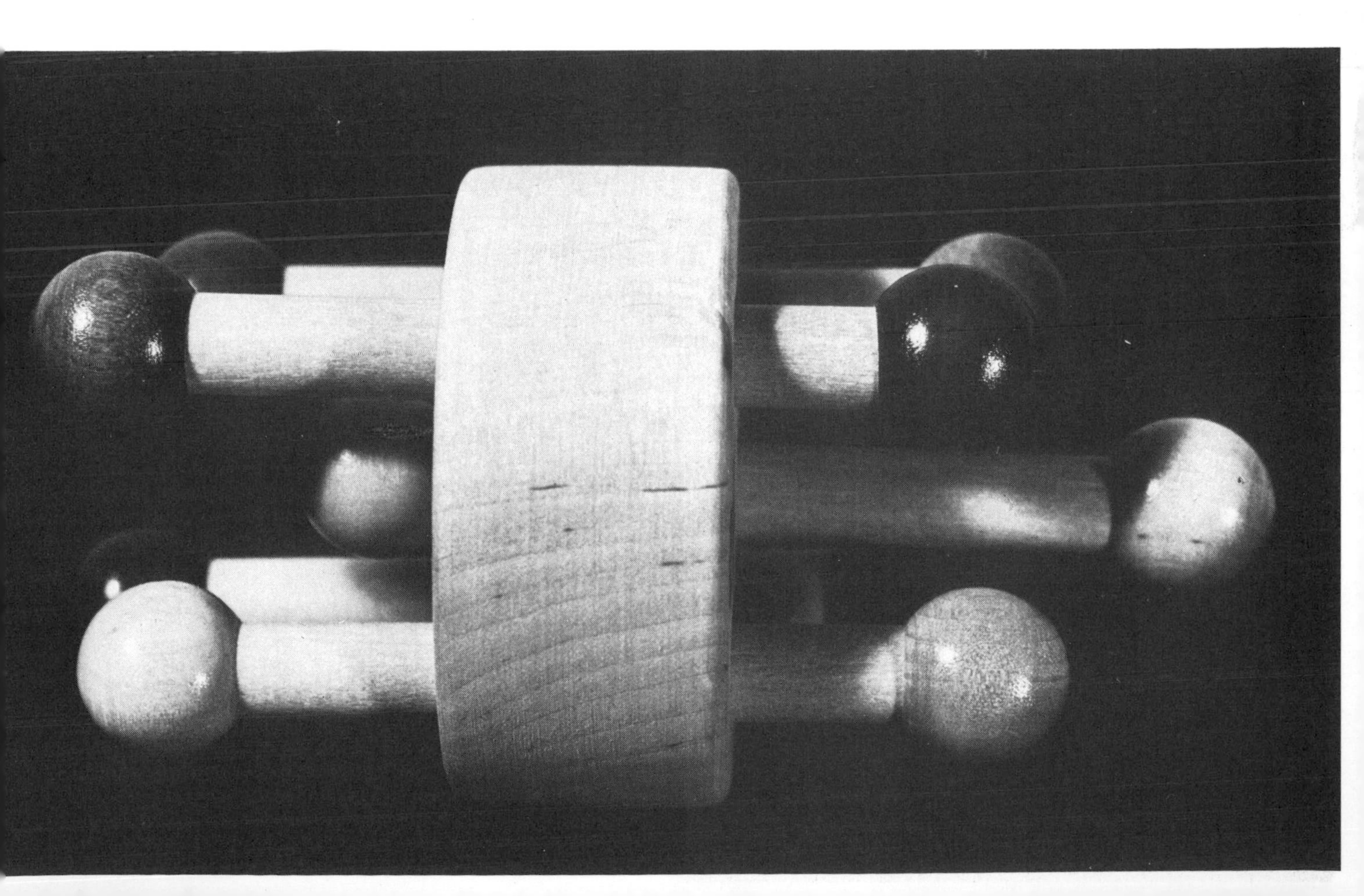

Fauni troll figures — "Papa Mumin" by Tove Jonsson, and Fauni figures by Helena Kuuskoski. I believe these troll figures fall into a special category of fantasy which exempts them from the normal evaluation one places on objects for children. These Fauni figures are handmade out of leather and fur scraps, and they are filled with either sawdust and glue, or pieces of styrene foam. The trolls are not deliberately intended as play toys — they symbolically represent the fantasy world of a rich Finnish and Scandinavian heritage. If the child wishes to use the troll as a toy, then he has gained an imaginative end beyond the symbolism. As long as childhood literature is filled with troll fantasy — then I believe these troll figures have their place as symbols. My own children very seldom "play" with them as one would play with a doll, but to take the troll away from them would be like cutting off their physical contact with a fantasy world they seem to thrive upon.

Children's bowl and cup by Peter Winqvist, and manufactured in ceramic by Arabia.
The idea behind this bowl form was to develop an interior lip at the top edge of the bowl rim. This lip was supposed to help the child in its first attempts at self-feeding. As the child brought a spoon filled with food up against the sides of the bowl, the overhanging lip was supposed to help the child keep the food in the spoon. Between the time this bowl was first conceived as a prototype, and the time it finally appeared as a manufactured product, the functional lip had all but disappeared. Compromise after compromise resulted during the process of bringing the form from conception to production. As the photographs indicate, the lip on the manufactured bowl is now no more than a figment of the imagination. As for the cup which belongs to the set — well, the photographs are self-evident. Mila, who tried this set out for the first time, spilled the cup two times during the first meal. There is no question that the cup is top heavy. Somehow it seems contradictory that a form deliberately conceived for small children should be manufactured in a breakable ceramic material, especially when the manufacturer also has

ccess to high quality plastics. If all small children can be expected to drop things, then why should the designer and the producer endorse the problem by providing breakable implements? One of the most interesting points about the set is that it is beautifully packaged as a mini-suitcase — perfect for a child to carry about and to stimulate curiosity. Mila had never before seen the box. Her delight is self-evident.

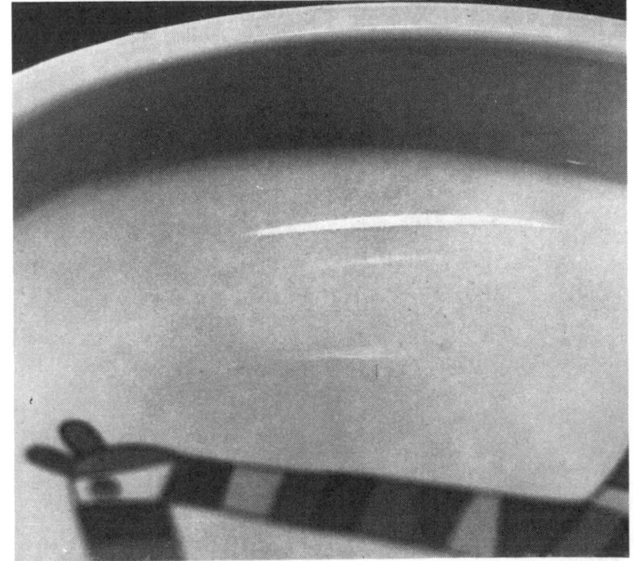

As a postscript to this form, I should like to record the way in which Arabia first advertised the form visually. They took a group of small children from the employee kindergarten, dressed them up in clean party clothes, arranged an "aesthetic" display of bowls and cups on one side of the table, and placed 2 or 3 children on the other side of the table with a spoon in one hand and a bowl in front of them. The bowl was of course empty. These mannequin children were photographed eating air. Arabia obviously did not want messy little "kids" spoiling their "aesthetic" piece of advertising. When this advertising appeared, it gave the impression, without any doubt, that these mannequin children were PROVING the validity of the form by eating out of it.

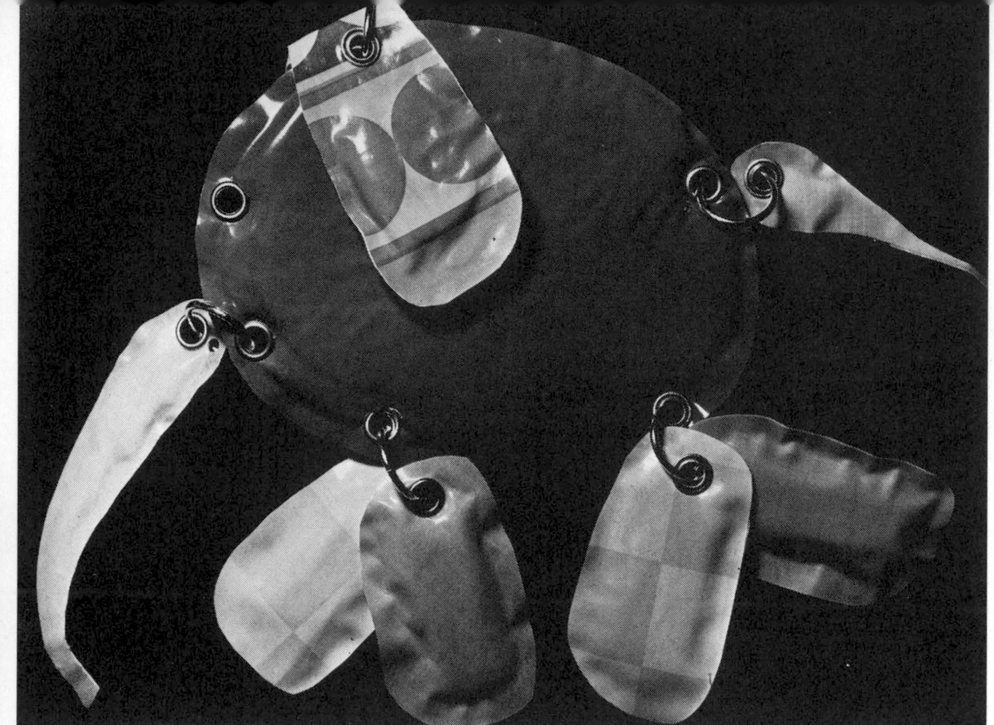

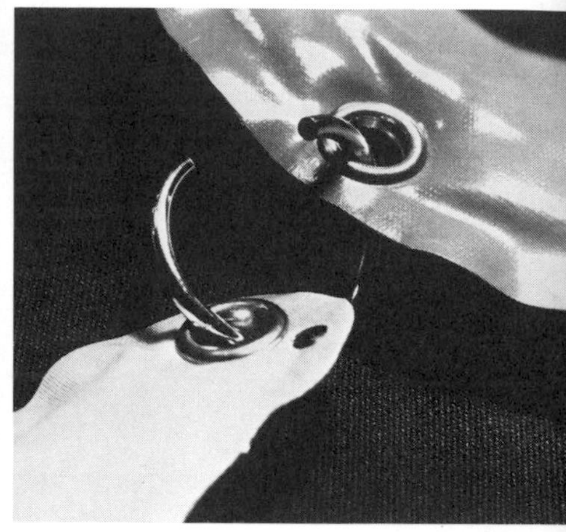

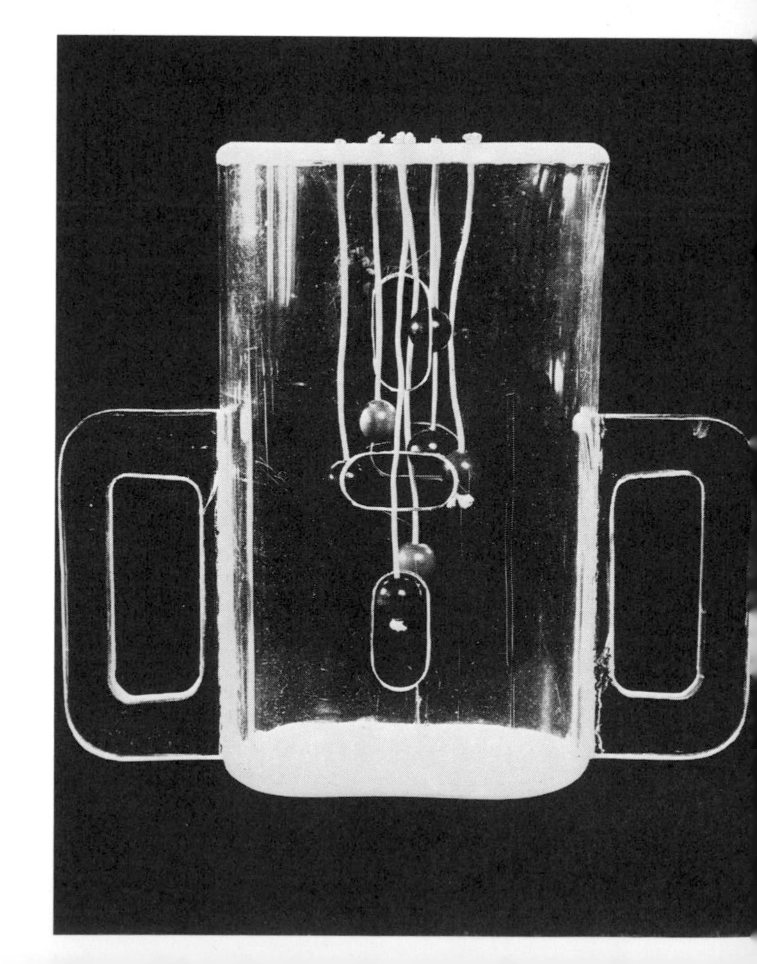

The two forms on this page were designed by Barbro Siltavuori. The forms resulted in response to the challenge Victor Papanek directed toward Finnish designers to concern themselves with "real" social inequities.
Both forms are prototypes for CP (cerebral palsy) toys, and are concieved to encourage muscular coordination. With the elephant, the child is supposed to open the rings in order to assemble and disassemble the parts.
With the transparent box, the child is supposed to learn to put his fingers inside the small holes in order to retrieve the colored beads. The beads are attached to the box with elastic thread and return to place once they have been retrieved. Mrs. Siltavuori presented these prototypes to several Finnish firms. None of them showed enough interest to manufacture them.

All terrain vehicle — it runs on land, on water, and on snow. In contrast to the single-mindedness of the American winter "Ski-doo", this vehicle offers the consumer all-season, all-terrain use. The technical construction was worked out by Olavi Karhu and Nils Fagerstedt— the form was designed by Antti Siltavuori. The vehicle runs on a 2-cycle engine of 12,5 hp with 6 000 revolutions per minute. Its total weight is only 250 kilos. The steering, acceleration, and brakes are worked from a single control lever. Optional rubber tracks mounted over the wheels make it a winter snow vehicle. Despite its mininum weight, the frame structure is made from steel and aluminum plate. This vehicle is still in its prototype stage. It has not been manufactured. Will it find a home in Finnish industry, will it become an exported idea, or will it die? If modern man can afford the luxury of an expensive, single-function "Ski-doo", can he also afford the convenience of a rational, reasonably priced, multi-functional form of Finnish design?

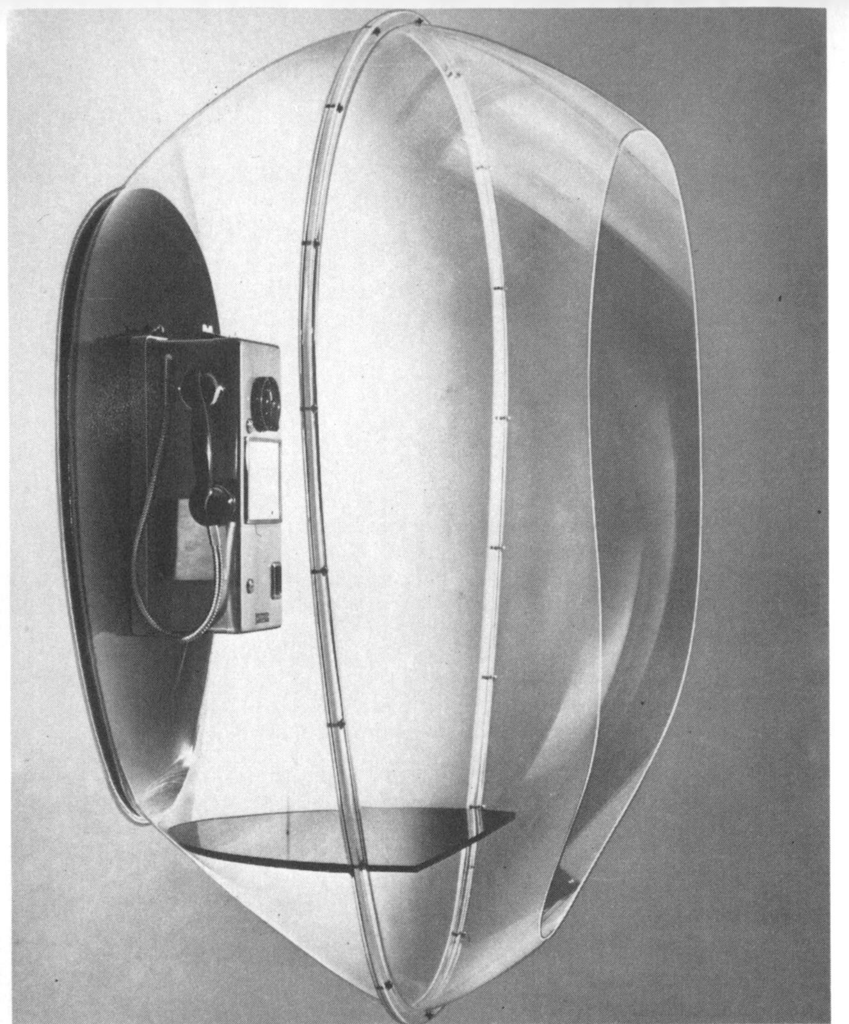

Polyplex telephone booth by Oy Polykem Ab. This telephone booth is manufactured in both a transpare and smoke-green acrylic plastic. It is a two-piece construction and comes equipped with a telephone book shelf. The manufacturer recommends the forr for use in public places such as banks, hospitals, and restaurants. Although these forms are admitted "aesthetic", I have begun to seriously doubt whether i fact they are a functional improvement over the older, self-enclosed square booth. The purpose of using a telephone in a public place — used to be to talk with somebody else on the telephone rather than to lo upon and appreciate the "aesthetics" of the telephone booth. Almost one-third of the surface area on this Polyplex booth is permanently exposed to room noises. In photographing this booth in restaurants, the Helsinki tunnel, and even in several airports, we learned a lot about its shortcomings. First of all, it is difficult for telephone users to clearly distinguish the edges of the booth opening. The transparent acrylic plastic, especially up against a window or a light colored wall, makes the opening difficult for ma people to see. Many people must feel the edges with their hands in order not to knock themselves in the head. Still another major disadvantage is that the womb-like form is limiting. The manufacturer may well intend that it be placed high on the wall, b in many of the installations we have encountered, t booth was placed so low that a telephone user had duck his head to get in, or to simply stand outside and try and stretch his arms to reach the telephone. This form could well go back onto the drawing board to improve upon these basic faults. When a form represe itself as public, then in fact it should represent the ultimate in open-endedness, rather than a basically closed and limiting form.

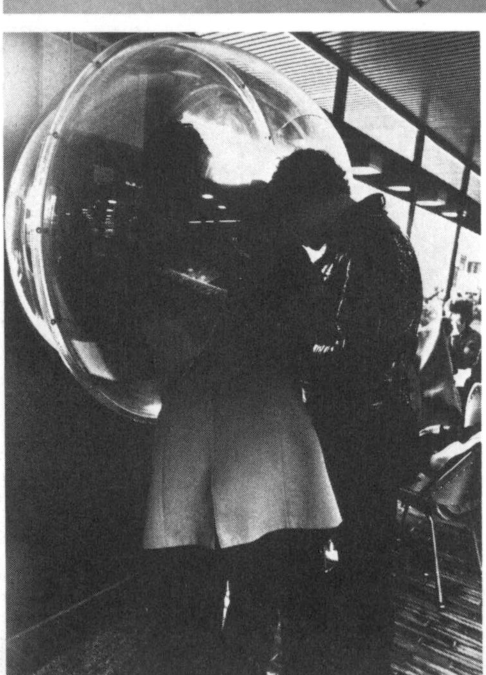

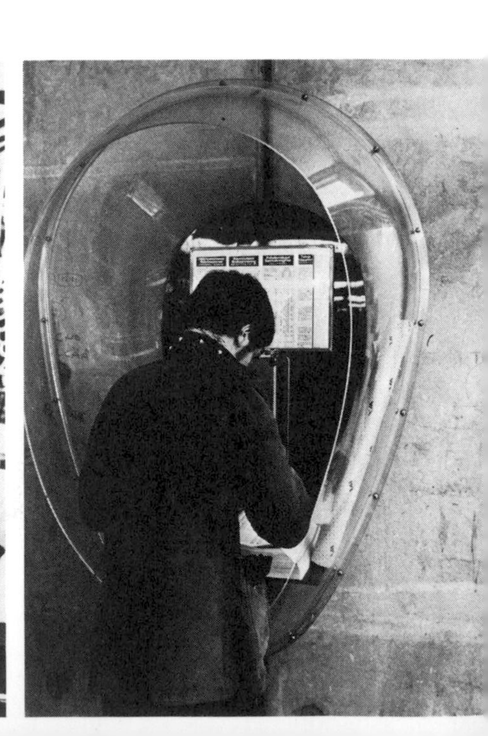

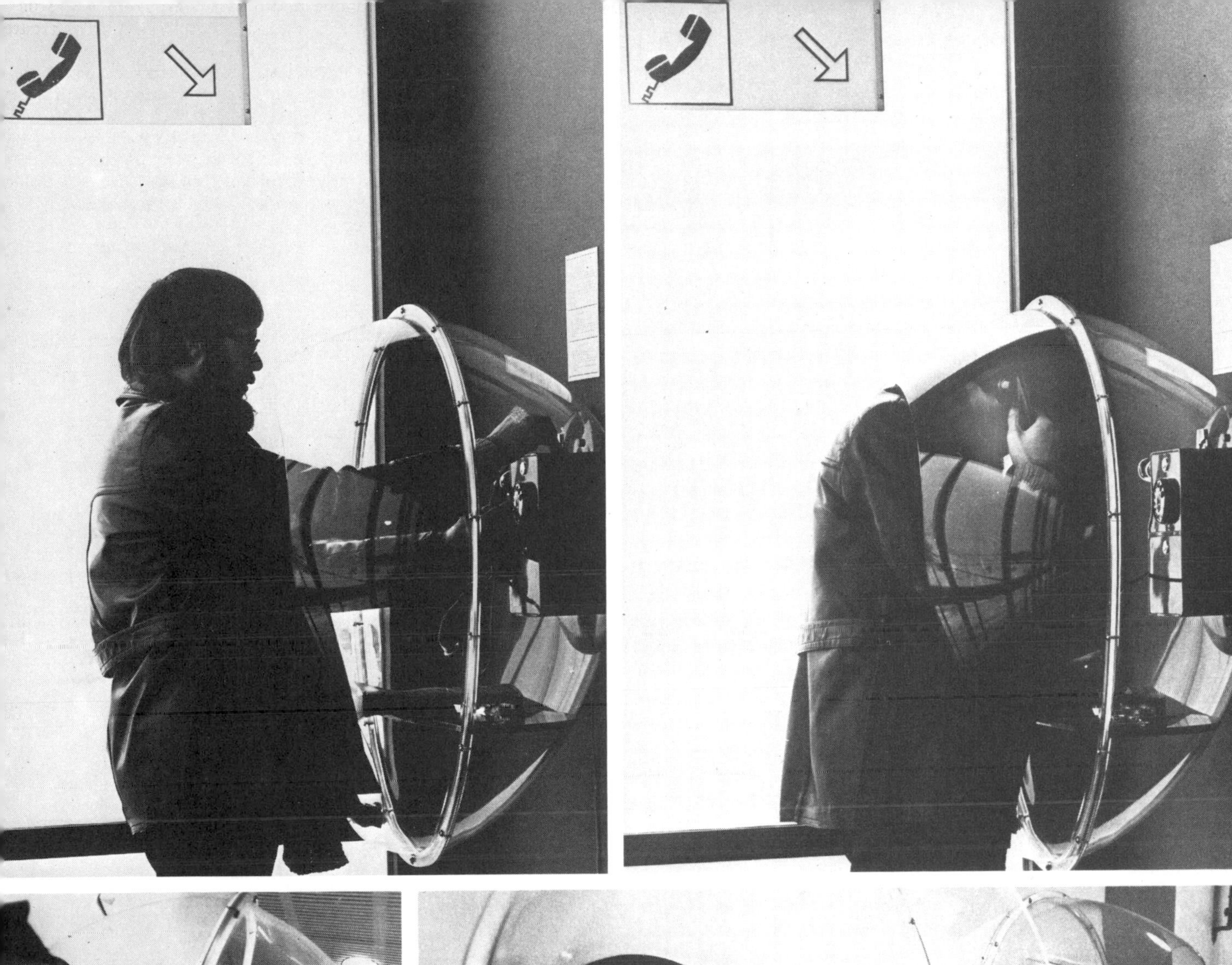

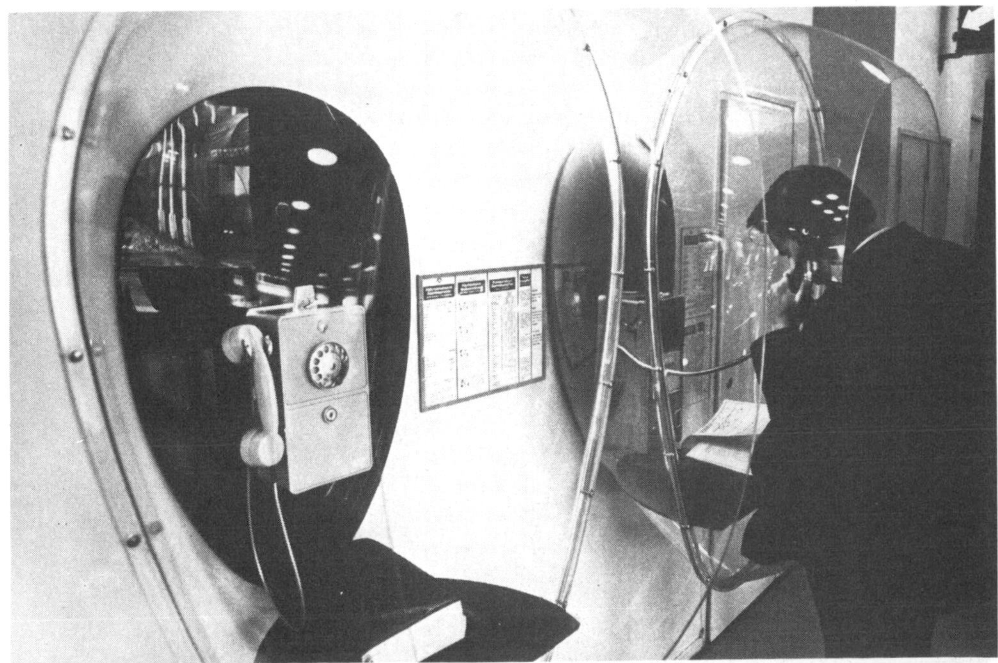

Stainless steel industrial scooter, manufactured by T:mi Erkkola. The scooter is fabricated for industrial and business use — it is not intended for the average family budget, although it can easily be driven by children. The form is produced by a firm which manufactures folding wheel-chairs, and both motor-driven and hand-driven invalid vehicles. As illustrated in the photographs from the Helsinki airport and from the Finnhand exhibition center, the form is intended for use in large buildings, or even large industrial complexes where people must walk great distances in order to communicate in their working environment. The newest model of the scooter is equipped with two back wheels and one front wheel — making it extremely stable, and well balanced. It is equipped with a front hand break, a bell, and an optional basket for carrying files and books. The rubber tread tires makes it run quietly indoors, and durable enough to use outdoors. The form is so well constructed that it looks as though it could outlive its owners.

HISS
LIFT
2. ND FLOOR / DEPARTURES
1. KERROS / SAAPUVA LIIKENNE
1. VÅNINGEN / ANKOMMANDE TRAFIK
1. ST FLOOR / ARRIVALS

Gasoline station, manufactured by Oy Polykem Ab. To my mind, one of the uglies architectural scars of the twentieth century has been the drowning of populated landscape with gasoline stations. For example, in America, the problem is out of control — it i common to find at least four competitive gasoline stations at every major city intersection or countryside crossroad. The problem is recently noticeable also in Finland. These Polykem gasoline stations begin to offer an alternative form to the structures which prey lik locusts upon the landscape and the people. It is unfortunate to note, however, that within its structural concept, Polykem seems to hav done nothing about controlling the gaudy advertising signs in these same gasoline stations. Although Gulf expects to install these stations, the photographs indicate that neithe Gulf nor Polykem has done anything about controlling this super-sell advertising. One therefore begins to wonder if the form is conceptually an alternative, or if it is simply a gimmick to sell even harder.

Futuro, manufactured by Oy Polykem Ab. This form, which has even made the pages of "Playboy" magazine, has come under a great deal of internal criticism by a number of Finnish architects. Admittedly, the interior curved surfaces present a number of practical problems for the owner, and the portability of the form presents problems in the dependence of the form upon earthbound facilities such as water, sewage, and heat. Otherwise, it is my feeling, despite negative criticism, that the form begins to offer an alternative to a perpetual landscape of rectangles and squares. As can be seen from the illustrations, the International Promoting Company in Bruxelles is in the process of constructing an entire hotel made up of stacked Futuro units around a center service cylinder. Is it not possible that functional structures can be defined in terms other than endless kilometers of grey squares and rectangles — the forms we now define as exclusively "modern" architecture?

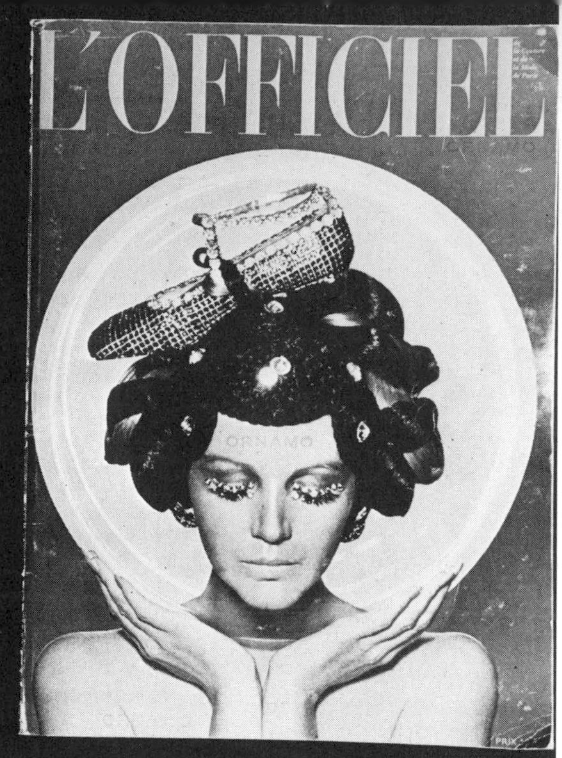

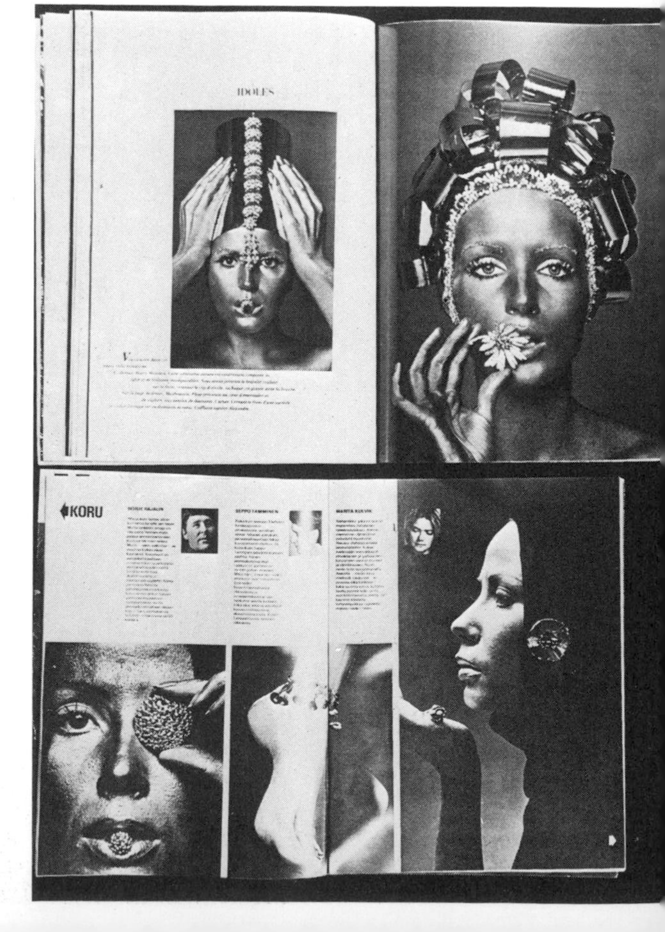

Where do glamour ideas really come from? These photographs illustrate a comparison between two fashion magazines — "Hopeapeili" of Finland, and "L'Officiel" of France. Both magazines attempt to glamourize jewelry — to make it photographically look better than it really is — by strapping it across the greased faces of fashion mannequins. Was the "Hopeapeili" article a reflection of Finland, or was it an import from the make-believe sales "genius" of Paris?

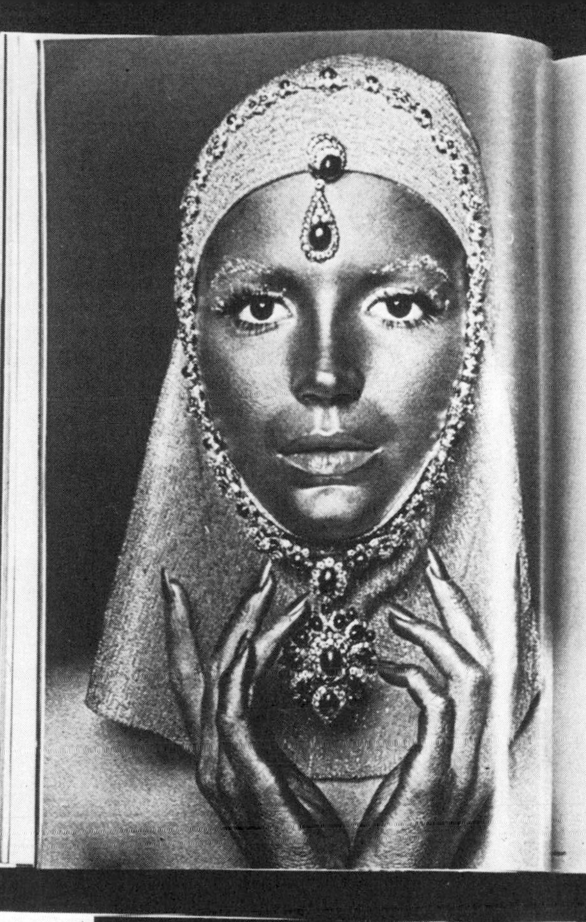

IDOLES

KORU

PENTTI SARPANEVA

BENGT ERIKSON

TAPIO WIRKKALA

FINLAND
It seems that nowhere else
in the world are the flowers
of high summer so luxuriant,
so fresh and so enticing
in their abundance
as in Finland.

It see
in the
of hig
so fre
in the
as in

at nowhere else
ld are the flowers
nmer so luxuriant,
nd so enticing
undance
nd.

marime

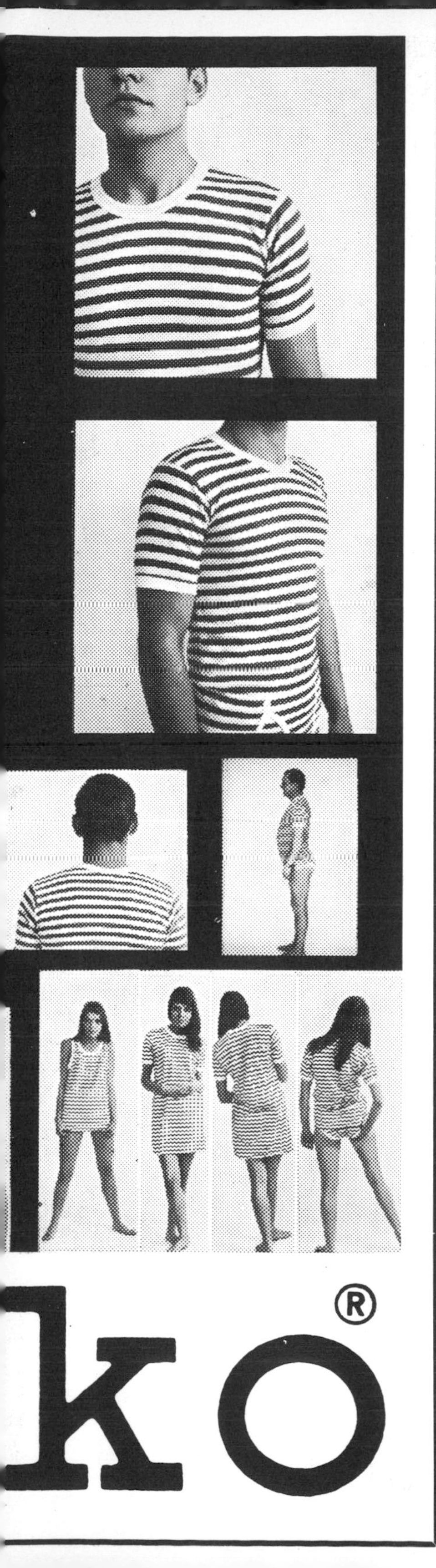

Honest advertising? This is the Marimekko advertisement that appeared shortly after the firm introduced its cotton-jersey underwear and pajamas. The layout was made by Timo Liipasti. Mostly, the advertisement is just people — fat, thin, long, and short. The non-text visual is strong, it is graphically exciting, it supports an excellent product, and the combination is honest. It is possible!

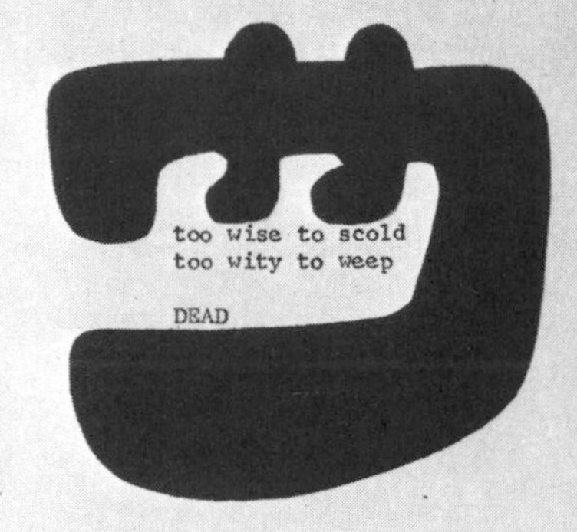

too wise to scold
too wity to weep
DEAD

I believe the photographs on these two pages represent a taste of the non-limiting optimism of youthful energy in Finland. Daniel Chompré, together with his Finnish wife Tuulikki, and weaver François Garrat have explored the imagery of the "glyph". Mr. Chompré, a French graphic artist, has been living and working in Finland for several years. He expresses his imagery in terms of the timeless "glyph", and rejects the concept of "art" as a medium for only museums and intellectuals. As the photographs indicate, Mr. Chompré has taken

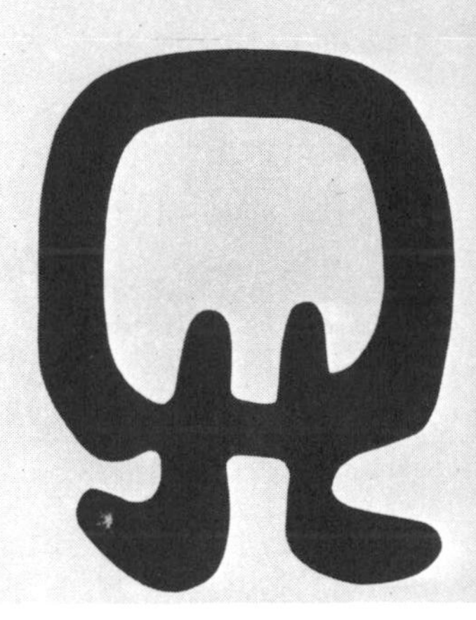

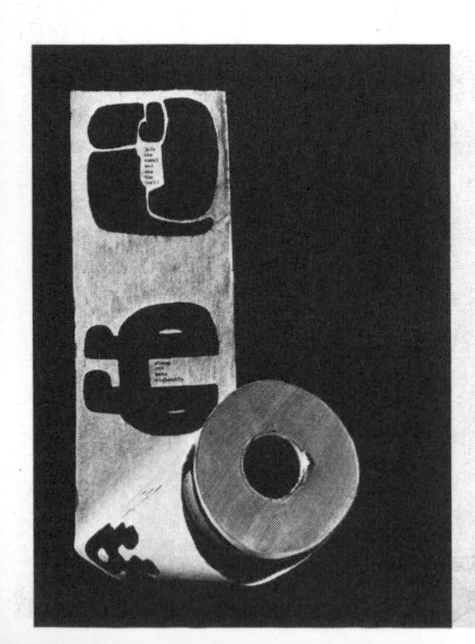

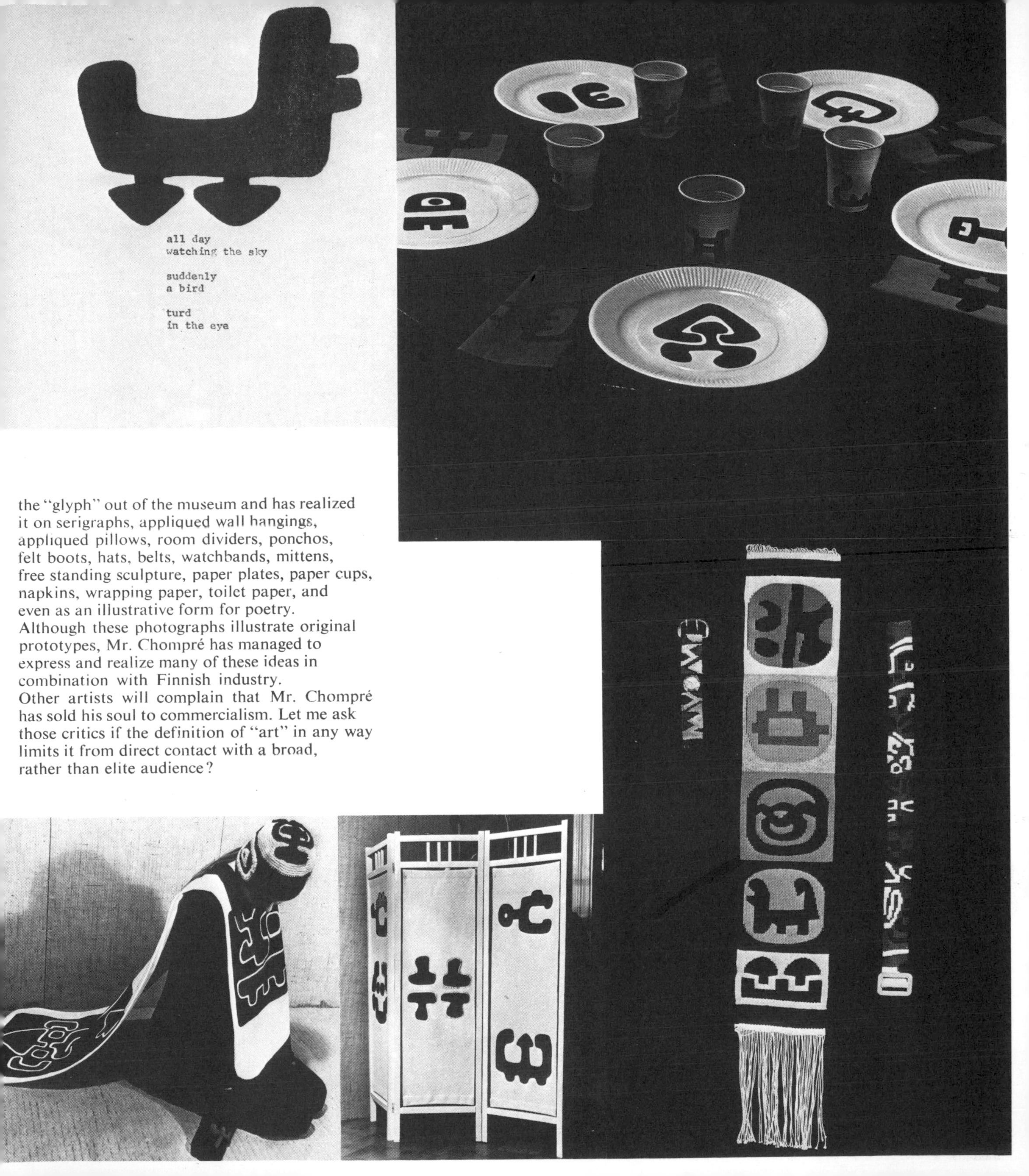
all day
watching the sky
suddenly
a bird
turd
in the eye

the "glyph" out of the museum and has realized it on serigraphs, appliqued wall hangings, appliqued pillows, room dividers, ponchos, felt boots, hats, belts, watchbands, mittens, free standing sculpture, paper plates, paper cups, napkins, wrapping paper, toilet paper, and even as an illustrative form for poetry.
Although these photographs illustrate original prototypes, Mr. Chompré has managed to express and realize many of these ideas in combination with Finnish industry.
Other artists will complain that Mr. Chompré has sold his soul to commercialism. Let me ask those critics if the definition of "art" in any way limits it from direct contact with a broad, rather than elite audience?

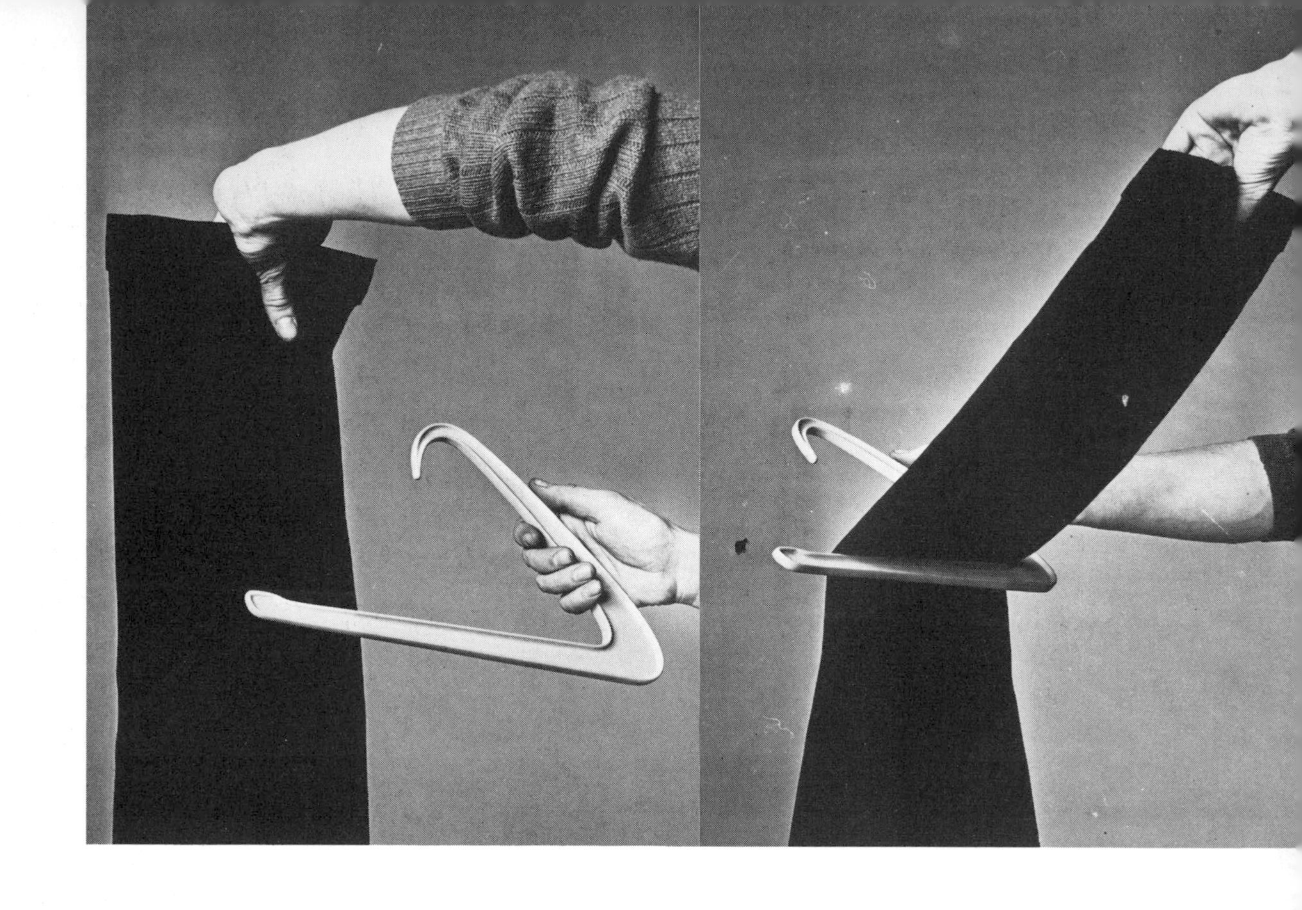

Raimo Nikkanen, a second year student at the Ateneum, was given a classroom project to re-design an ordinary pants hanger. This plastic prototype was his answer — convenient, cheap, and attractive. Did somebody say there was no youthful talent in Finland?

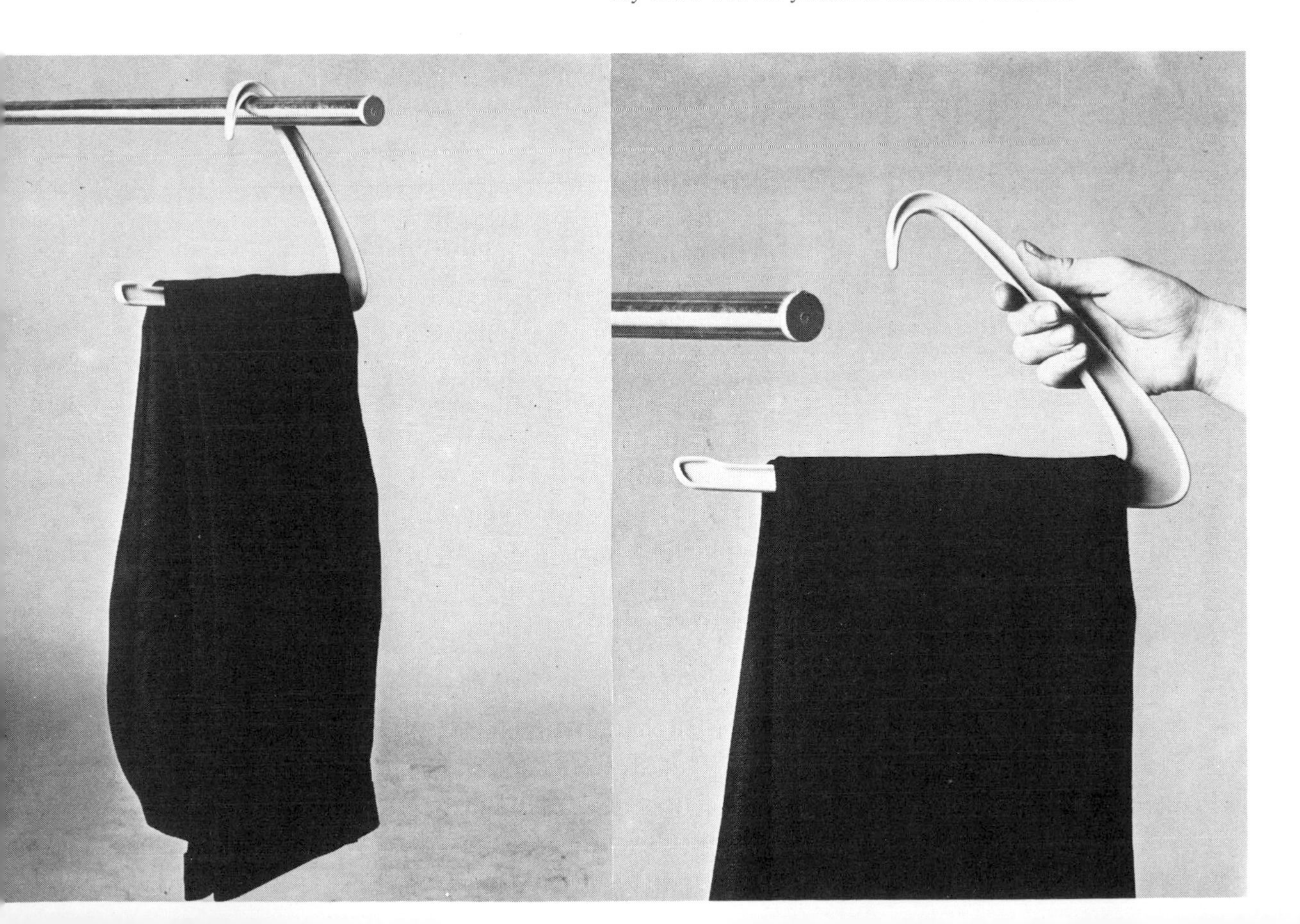

The cape and poncho illustrated on these two pages were designed by Jukka Vesterinen. They represent still another sampling of the unexplored youthful talent existing in Finland. Mr. Vesterinen began working in one of the large Finnish textile factories as a "designer", but he was all too quickly bored by the conservatism and the absence of challenge. He felt his energies were being unused (unfortunately a very common story). He consequently left his position, and began a small shop of his own — making garments with a knitting, and sewing machine. In his case, he felt that this move away from industry was his only insurance against ending up a perpetual flower artist. Does Mr. Vesterinen represent one spark of a potential fire that needs anticipating before, rather than after it has exploded?

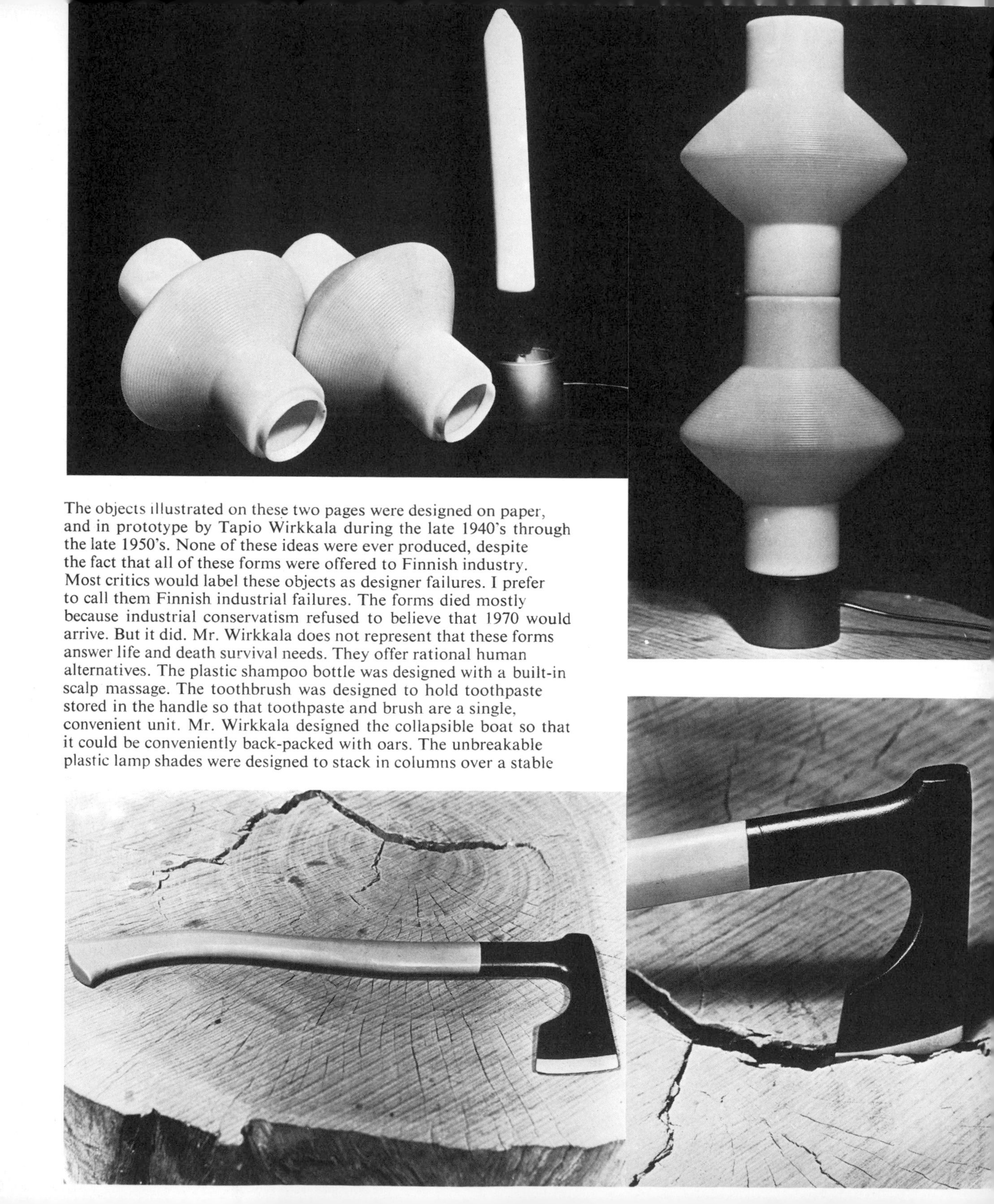

The objects illustrated on these two pages were designed on paper, and in prototype by Tapio Wirkkala during the late 1940's through the late 1950's. None of these ideas were ever produced, despite the fact that all of these forms were offered to Finnish industry. Most critics would label these objects as designer failures. I prefer to call them Finnish industrial failures. The forms died mostly because industrial conservatism refused to believe that 1970 would arrive. But it did. Mr. Wirkkala does not represent that these forms answer life and death survival needs. They offer rational human alternatives. The plastic shampoo bottle was designed with a built-in scalp massage. The toothbrush was designed to hold toothpaste stored in the handle so that toothpaste and brush are a single, convenient unit. Mr. Wirkkala designed the collapsible boat so that it could be conveniently back-packed with oars. The unbreakable plastic lamp shades were designed to stack in columns over a stable

on base — simple construction at low cost. The woodsman's
e was designed to meet a number of problems encountered with
aditional axe forms. Mr. Wirkkala's axe can be held in the hand
nveniently for shaving bark for starting fires; it has a fibre-glass,
on-breakable handle molded into the blade so that the axe head
n never fly off; it has a hammer pounding surface; and the form
the blade conveniently fits over a man's belt for carrying. One
the several reasons for including these forms in this book is to
vidly illustrate to young Finnish designers that even their "father"
gures (who they now often accuse for having no social context).
ave also encountered rough industrial rejection for valid forms.
ow, ten years after (and ten years too late for Finland), several
Mr. Wirkkala's ideas have been realized by other
ountries and other non-Finnish designers.

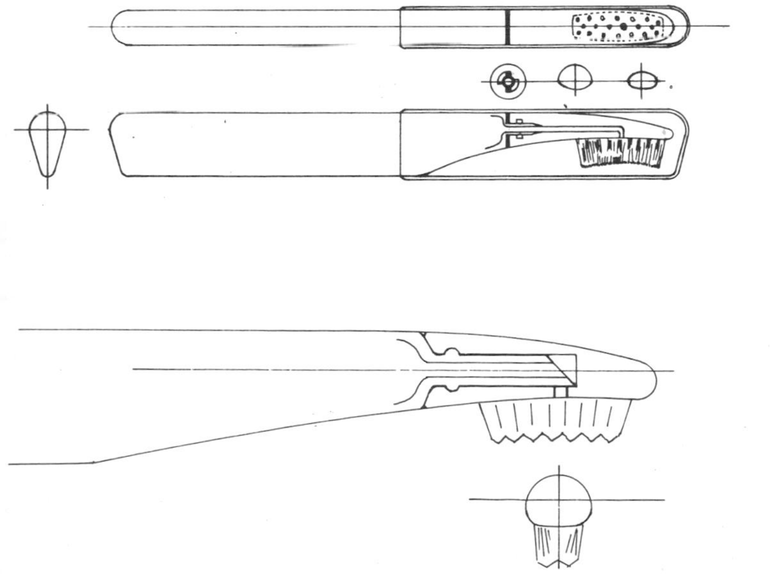

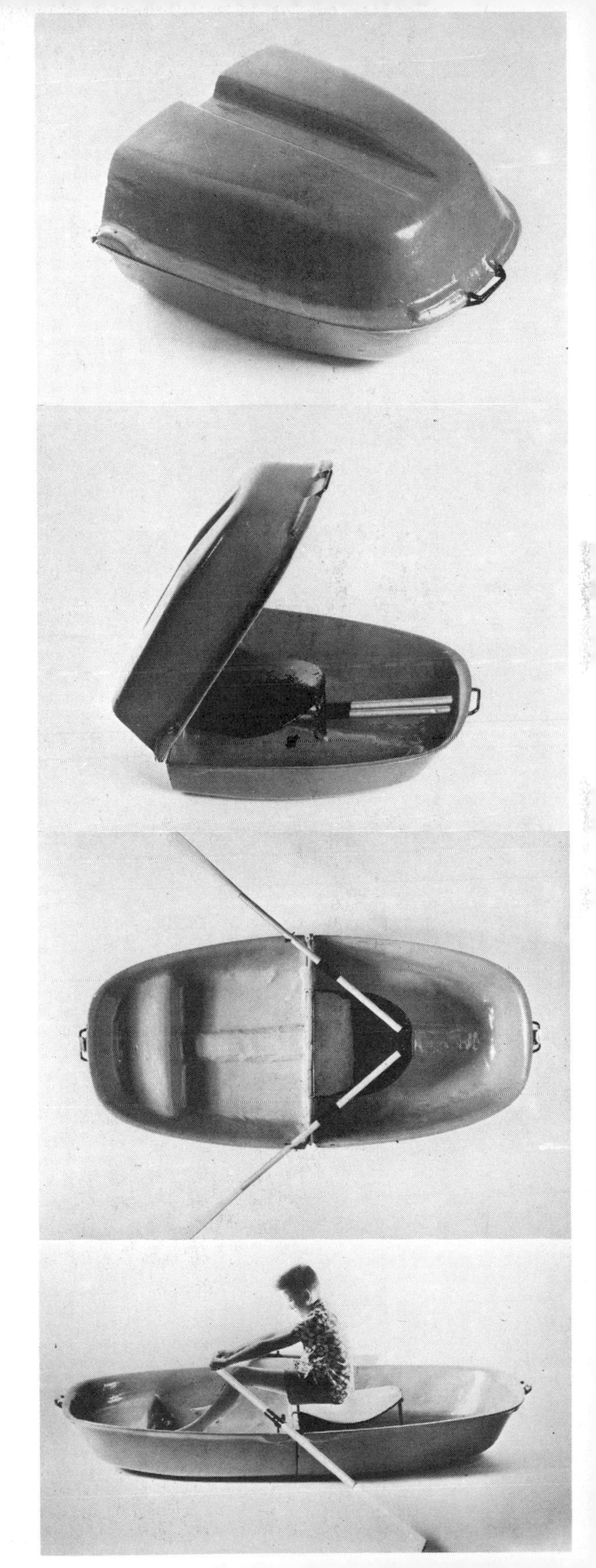

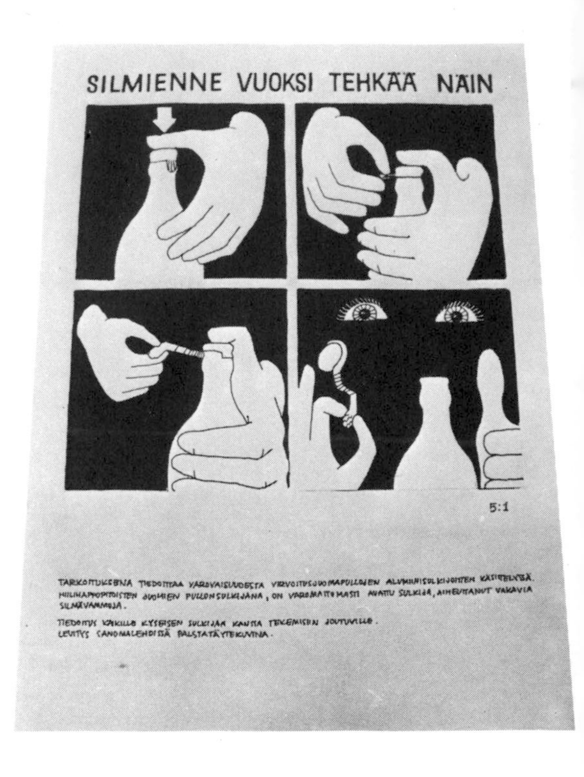

Comments by Ateneum students on a few of the urgent social and objective problems confronting the Finland of the 1970's, including air pollution, wildlife pollution, the accumulation and disposal of refuse, as well as graphic comments on our over-indulgence in glamour, and the infinitive lengthening of man's so-called "basic material necessities". The student energy represented within this graphic concern must be invited and then supported in an attempt to begin solving these problems. We have all looked — and talked — much too long. An idealism that concerns itself with our environment and our fellow man — MUST NOT BECOME ONLY A CLASSROOM EXERCISE IN THEORY. General and specific improvements can be made if the energy behind these graphic comments is put to use to creatively participate in solutions.

The Place Of Design

A curious thing happens when you begin talking to people on the street, or people sitting next to you on a bus, or just talking to people anywhere and everywhere — about the subject of design. Ask them what they think about design and the designer, and then just listen — and learn.

"Oh, design? Isn't that something glass blowers do?"

"A designer? Aren't they the people who draw shapes for things like cup handles, frying pans, bathubs, and automobiles?"

"Design? Well, I'm not awfully sure, but I think it has something to do with objects".

An even more curious thing happens when you ask a designer what he thinks about his own profession, and specially his place in society. More often than not, he either doesn't know, or he will confirm what the man sitting next to you on the bus said about design being the profession for determining form. Very seldom will the designer identify his profession within a social context. On the contrary, he quite often feels that his is a kind of asocial specialty existing way out in the suburbs of society. Far too often he looks upon himself as a man who se problems begin and end with *objects*, rather than whose problems begin and end with *people*.

Yes, there have been many designers who have more or less divorced themselves from society into a kind of nearsighted objects clique, but they are not alone responsible for carrying out their asocial role. For the most part it has been society who has defined the role the designer has assumed. Yes, you and I — just everyday people — we are the ones who have misused the potentials of the designer. We have taken some of our most sensitive and creative human raw material, we have given it special professional training, and then we have cast it out of the bloodstream of out social context and into a specialty reserved only for objects. We have severely limited the creative energy that could have been one of mankind's closest links with itself. We have employed politicians, preachers, philosophers, social scientists, and other miscellaneous do-gooders to look after us. We have listened to their messages and we have acted upon their advice. And while we were doing all this, we have never even asked our designers what they thought, how they felt, or what they would recommend. We have given all the power to the politician, and we have given most of the intellectual honor to the philosopher, the preacher, and the social scientist — but we have never even listened to the man in whose sensitivities we have invested our objects.

We people are curious.

We have tried it the political way, the philosophical way, the religious way, and the scientific way, but we have not even had the curiosity to consult our designers.

We people are curious.

We have taken a man whose talents we have not yet begun to explore, and we have set him up to worry about the shape of our cup handles. Should we not begin to ask him about the shape of our lives? Should we not begin to listen and contemplate his advice on the shape of our future? Should we not begin to break down the limitations we have set upon him, and should we not invite him in from the suburbs to participate aggressively in our social context?

We people are curious.

We criticize the designer for not involving himself in our social lives, but then we take away his opportunity. We set up awards, foundations, and prizes based upon social idealism. We invite the designer to participate. He suggests a rational means of controlling the destruction of our natural environment, or a way to solve some of our social problems, and then what do we do? We do exactly as was recently done in West Germany at an international competition encouraging design idealism. We give our support, our money, and our awards to another ingenious fellow who has managed to figure out a device for cutting off the top of an egg. An absurd egg slicer takes preference over a rational idealism which can help all of

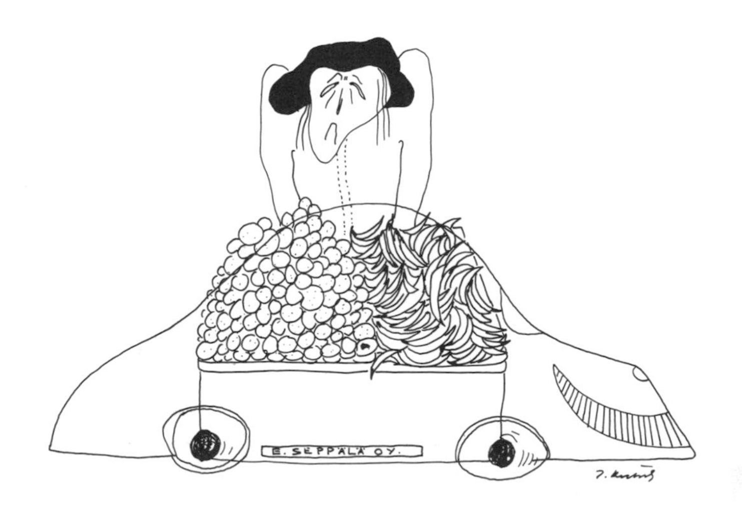

us save ourselves. We reward frivolity and ignore survival.

We people are curious.

You know we really don't need to go to underdeveloped places to begin saving ourselves. We can begin right in Finland, or wherever we call home. First of all, we can recognize that hundreds of young Finnish designers, just like thousands of young designers all around the world, have re-established contact with their social conscience. These people feel a need to do something beyond constructing cup handles and frying pans. They live in a world of social problems and social inequities, and they have the creative talent and energy to realize something positive. We can let them help us help ourselves by simply giving them an opportunity. There is no reason for them to go running off to so-called underdeveloped nations just to fulfill their idealism when we can use all of this energy right on our own doorstep for yet several generations into the future. We can invite them to solve some of the absurd social messes that exist right in Finland.

We people are curious.

We tell our designers and architects what to do instead of asking them to find rational solutions. For example, we totally ignore the lack of public facilities for bus and train travelers in Finland even though almost all of us use these facilities by the hundreds of thousands every day of our lives. What do we do instead? We employ our designers and our architects and we tell them to build airports, but we don't want just ordinary airports. No, we must have BEAUTIFUL airports. The air traveler must be treated to luxury while the bus and train traveler (you and I) are treated like animals in a farmyard. We build half empty aesthetic monuments in places like Turku, Pori, Helsinki, and elsewhere in Finland, but only a small percentage of us ever get inside them. Instead, the overwhelming majority of us who daily travel on trains and busses prefer to stand out in the cold, or sit in dirty and overcrowded waiting rooms drinking lousy coffee out of paper cups. We do not consider this a designer problem.

We people are curious.

We have lost our overview by becoming non-communicating specialists. We seem to go forward by proceeding backward. We define the solution to our problems even before we employ the designers and architects to help us. We have defined the task of both designer and the archtitect in terms of objects. We believe that it is their responsibility to make objects to solve all of our problems. We demand objects from them in every situation, but fail to realize that more than half of our problems may not even require object solutions. Why do we limit the use of our most sensitive and creative people? Objects are not the only result of sensitivity and creativity. Perhaps if we give our designers and architects a little freedom, and stop limiting them to solving all problems with objects, we can make use of their energy so they can realize the overview we ourselves have lost.

We people are curious.

We gather a group of businessmen together, add a dash of politics, and a pinch of city planning — and we come up with a recipe for a Helsinki walking street that violates all common sense. We have decided to be contemporary and "groovy" urban Finns with our very own walking street just like in Copenhagen, but we totally miss the point of a walking street, and several of our citizens have suffered personal injury because of our ignorance. We never bother to ask our designers and our architects to inform us about the CONCEPTS of pedestrian streets. No, we do it ourselves. It's as simple as baking a blueberry pie. We just simply call Aleksanterinkatu a walking street. Now doesn't it sound just like Copenhagen? I like the ring to the English words, don't you? We don't give a tinker's damn about concepts and pedestrians. No, we run trams down the middle of our walking street like savages — we threaten the life of every citizen who tries to walk in our walking street, and we end up with a walking street that is not a walking street because everybody walks on the sidewalk unless he's crazy. And the funny thing is, we all profit off of our miserable failure. The businessmen have made a killing getting rid of garbage, and the urbanite has enjoyed the thought that his little Helsinki is now so international that it has its very own walking street. We enjoy our mess and refuse to think that pedestrian problems are designer problems.

We people are curious.

We employ our architects to build fantastic new churches, and we even blow thousands of tons of solid granite into bits just to bury the churches aesthetically into the earth. While the architects are building the outsides, we employ textile designers and interior designers to furnish these churches with a total blindness for economy. We give them license to experience a creative orgy, and we pay the bill, thank them, and are well pleased with the results. But hardly any of us use these churches. Many of them stand half empty in the shadow of already existing half empty churches just a few hundred meters away. Our churches make fine monuments for tourists to photograph and to bring friends into when we want to show them our "exotic architectural flair", but almost none of us can economically afford to perpetuate this kind of irrational spending.

We people are curious.

We direct our architectural and designer talent into churches, while in contrast, and just outside the churches, we suffer from a national alcoholic problem that has all but dumbed our senses. We tell our designers to create tapestries, lighting fixtures, furniture, and other religious equipment for the insides of our churches, while outside there are thousands upon thousands of homeless Finnish men and women sleeping in garbage cans, eating from garbage cans, falling over on the streets around us, and freezing arms and legs, or going blind or even insane from the effects of wood alcohol. We see these tragic people every day of our lives and have grown all but immune to their suffering. One falls down in front of us in the city. Fifty people walk right by him without extending a hand. We see dozens of them invade the Helsinki market place every afternoon to pick through garbage for food scraps. We see one trying to flop himself into a garbage can for a place to sleep. He tries not to lose his beltless trousers. It is like watching a Charlie Chaplin movie. We think we are watching a humorous movie and are insensitive to the fact that this is a real human being. We see 15—20 homeless men sleeping on the cold ground pressed up against a city heat duct near Linnanmaki amusement park every night of the long winter and we do nothing about it. We see dozens of homeless men scattered about, as if their bodies were chunks of wood, along the tram tracks number 8 in Helsinki, or near the old cemetery on the way to Lauttasaari. Even those of us who try ignore them cannot ignore them because they are everywhere — in the railway stations, the city underground, on the benches in parks, at the harbors — and some of them even find their way into our apartment buildings and office building hallways just to keep warm. They are Finnish men and women just like we are, but they have a problem — a problem of wood alcohol,

cheap vodka, and social inequities within our culture. Sometimes we stuff them into institutions to sort of get rid of them, but then we rarely give them professional psychiatric help, and almost never investigate the causes which create their problems and our problems. Once in a while, when we do get a bit of pity in our hearts, we call up somebody who knows something about buildings and tell them to construct a place to keep these people in. We ask for an object, we need to solve the problem with an object. Do we ever consider that many problems cannot be solved with objects? Do we ever ask our designer talent to seriously put their creativity into digging up the causes of Finnish alcoholism and the suffering of homeless men? No, of course not! We say this is not design. Design is cup handles, and frying pans, and beautiful glass for American and German tourists. Do we ever put an army of energetic young designers, backed by a rational budget, into exploring with the technicians and chemists a way of making wood alcohol non-harmful? No, we use our sensitive and creative people to give us gadgets to slice off the top of our eggs. We do not consider alcohol a design problem.

We people are curious.

We employ politicians and city planners. They recommend that we convert the Helsinki marketplace into a parking lot, and we almost go along with their advice. We have something unique — we have the most beautiful outdoor marketplace in northern Europe, and we almost wipe it out in order to park automobiles. We almost extinguish our most priceless landmark in favor of motors, steel, and exhaust. We listen to our advisors and barely escape destroying the most vital human institution in our capital city. We are almost convinced that markets are "old fashioned", and that if we want our capital city to be "progressive" — then DOWN WITH MARKETS AND UP WITH AUTOMOBILES. We do not engage a group of designers and architects to seriously study our city parking problems until the problem is beating us over the head. We do not consider parking a designer problem.

We people are curious.

Year after year we continue to spread our cities out into the land. We clear patches of forest, and then erect new apartment communities for commuting suburbanites. This expansion goes on bit by bit and there seems to be no overview. We successfully move these urbanities out into the land, but then we somehow forget that in order for them to live, they must come back into the city. Whenever the problem of transportation reaches a panic point, we plug up the hole temporarily by adding a new bus route. Eventually, we have added dozens upon dozens of new bus routes where no bus routes had existed before. Busses seem to be the cheapest and quickest solution to everything, so we take the path of least resistance. One morning, several years later, we try to drive in or out of our capital city during rush hour and we discover that our streets are jammed beyond capacity with busses. We stand on Kaisanieminkatu in Helsinki at 5:00 o'clock in the afternoon of any working day and watch the unfortunate traffic policeman being half gassed to death by exhaust fumes from busses and cars. We look up the street and see an endless mass of confusion buried in a thick cloud of poison air — our air — our clean, pure Finnish air! We have been watching television for years and observing senseless air pollution in America and other places — and we have thanked God that it could not happen here. BUT IT HAS ALREADY HAPPENED, AND IT IS GETTING WORSE AT EVERY TICK OF THE CLOCK. We have assumed that trams are anti-progress vehicles, and we do not consider extending tram lines until it is entirely too late. We have followed the excellent advice of our city planners, politicians, and other wisemen, but we have never considered using our young designer energy because we do not consider public transportation a designer problem.

We people are curious.

We think we have solved the problem of how to care for the children of working mothers. We have encouraged private day care centers, but we have failed to back our solutions with money. Several dozen day care centers have popped up all around our capital city. This has opened up a new small business for many younger and middle-aged women. That is good. They care for our children in their own private apartments while we are at work. Everything moves forward with a smile until the landlord and the other tenants in the building begin to complain about the noise and confusion of 15 to 20 small children in the building during the day. And so the day care center politely moves itself out into the city parks for as many hours as possible. But it is damned cold in those parks during the winter. We see children and teachers huddled together, clapping hands to keep warm, and stomping feet to keep out the bitter cold. We see them grouped up against the walls of buildings trying hard to smile and make the best of it. We see them, and we tell ourselves that a taste of cold winter air is good for children, but we fail to recognize that these children are overextending a taste of cold winter air. We do not enjoy standing in the cold many hours ourselves, but it is quite all right for our children. We, in fact, give them no alternative. We have plugged another hole with a temporary solution. We do not employ our designers to investigate this problem because we do not consider children a designer problem.

We people are curious.

We do not consider bus and train stations a designer problem.

We do not consider walking streets a designer problem.

We do not consider alcohol a designer problem.

We do not consider homeless men a designer problem.

We do not consider parking facilities a designer problem.

We do not consider public transportation a designer problem.

We do not consider air pollution a designer problem.

We do not consider day care centers a designer problem.

We do not consider suicide a designer problem.

We do not consider old age a designer problem.

We do not consider benefits to widows as a designer problem.

We do not consider unemployment a designer problem.

We do not consider narcotics a designer problem.

We do not consider education a designer problem.

We do not consider social inequalities to gypsies and Lapps a designer problem.

We people are curious.

We consider egg slicers a designer problem.

We people are curious.

We have elected the American way.

These photographs vividly illustrate a pedestrian walking street where pedestrians walk on the sidewalk and trams monopolize the street. Pedestrians even stand waiting for traffic lights to tell them when they are free to cross their own street. The other solution, which was tried earlier in the Helsinki tunnel, was to shove human beings underground, and save the air above so that it could be filled with the exhaust of automobiles, busses, and trucks. Pedestrians are not designer problems?

KOFF
KOFF
KOFF

We people are curious. Public transportation
is not a designer problem?
Traffic is not a designer problem?

HELKO
ASI-97
HILDEN
PUERTO
BHB 37

These pages illustrate the human pulse of the Helsinki marketplace along the waterfront. Will it one day fall as a victim of irrational modernization just in order to park automobiles? Is it not possible that designers and architects can begin NOW to find rational solutions to problems created by automobiles?

Without looking very far, and on an average spring day, we candidly snapped these 4 pages of photographs in Helsinki — scenes which are all too familiar to all of us. Somebody near the old cemetery has been kind enough to install a free water faucet. Is one free water faucet, sidewalk space to sleep on, and an occasional under-budgeted, and under-staffed institution the answer for Finland's problems of alcoholism and homelessness?

Chiquita

SANNA
KAUNEUSHOITOLA

This single photograph was taken in the Pori airport during the middle of an average weekday. Nothing but the best — Haimi chairs by Yrjö Kukkapuro, a fresh flower on each table, a whole glass observation wall for customer entertainment — all of this luxury for just one customer and one photographer. Now contrast this to what you already know about, for example, the customer conveniences in a public bus or railway station. Is there no room for designer improvement?

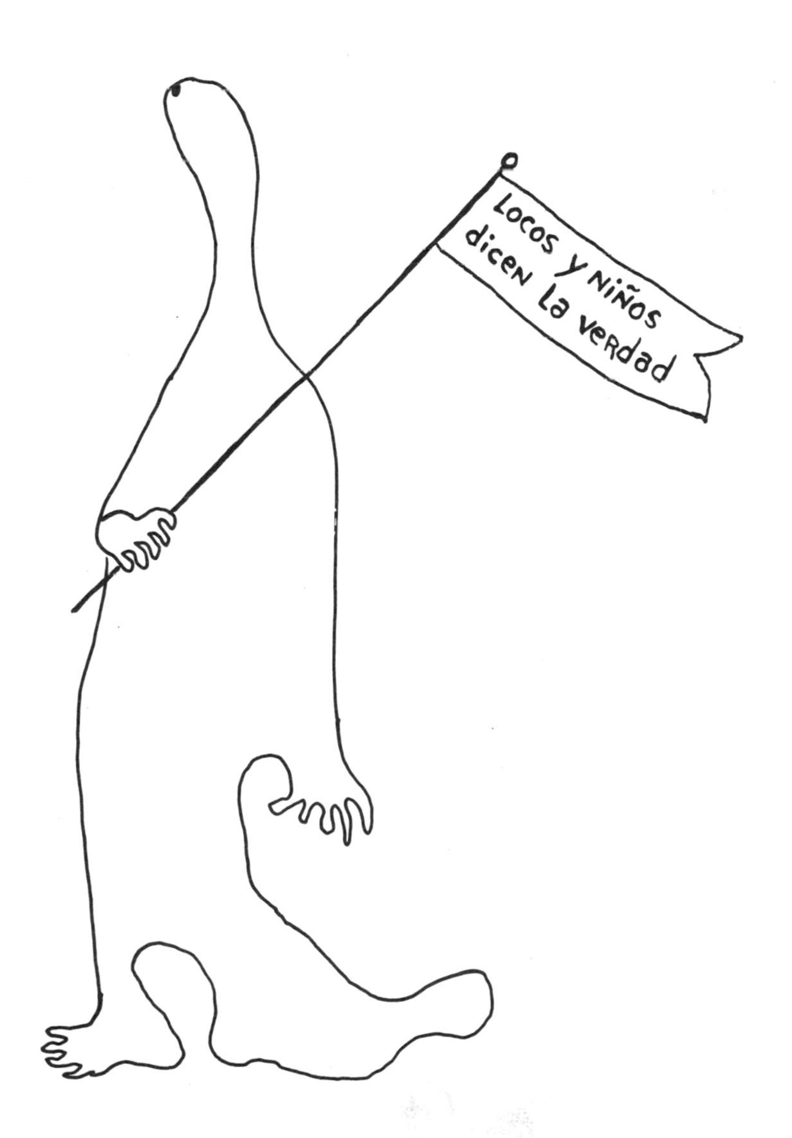
Locos y niños
dicen la verdad

Acknowledgments

A great many people have helped to build this book, and have helped to educate, enlighten, and endure its author. I thank you all, both individually and collectively. And thank you also to: Sinikka Salokorpi, Juhani Jaskari, Kari Haavisto, Juhani Pallasmaa, Daniel Chompé, Patrick Degommier, Tapio Wirkkala, Barbro and Antti Siltavuori, Anja Jaatinen, Martti Vuorenjuuri, Martti Berger, Maire Waldén, Kyllikki Salmenhaara, Päivi Harmia, Teemu Lipasti, Juhani Konttinen, Anna Tauriala, Simo Rista, Sakari Nenne, Cecile Numers, Agneta Bäckström, Olli Alho, Pekka Suhonen, Anto Leikola, Matti Timola, Peter Winqvist, and the photo laboratory at »Avotakka» magazine. Finally, a very special thank you to Inger Kulvik who's satirical gift with pen and ink have helped to hold all my rambling insanities together.

And thank you Finland!

About The Photographer

Ove Hector is a two-meter high Danish psychologist-photographer. He spent many weeks in Finland gathering the photographs for this book. Mostly, he travelled around on a motorscooter, with an author hanging desperately on the back. It was no easy trick to pack two grown men, three cameras, and about two dozen other photographic tools onto a motorscooter. Ove did it brilliantly.

Ove Hector has held one photography exhibition in Copenhagen, and he has contributed photographs to a number of publications both in Scandinavia and in the United States. When he is not looking like a Viking, he lives quietly in Copenhagen with his wife, Evelyn, and small son, Jakob.

Donald J. Willcox

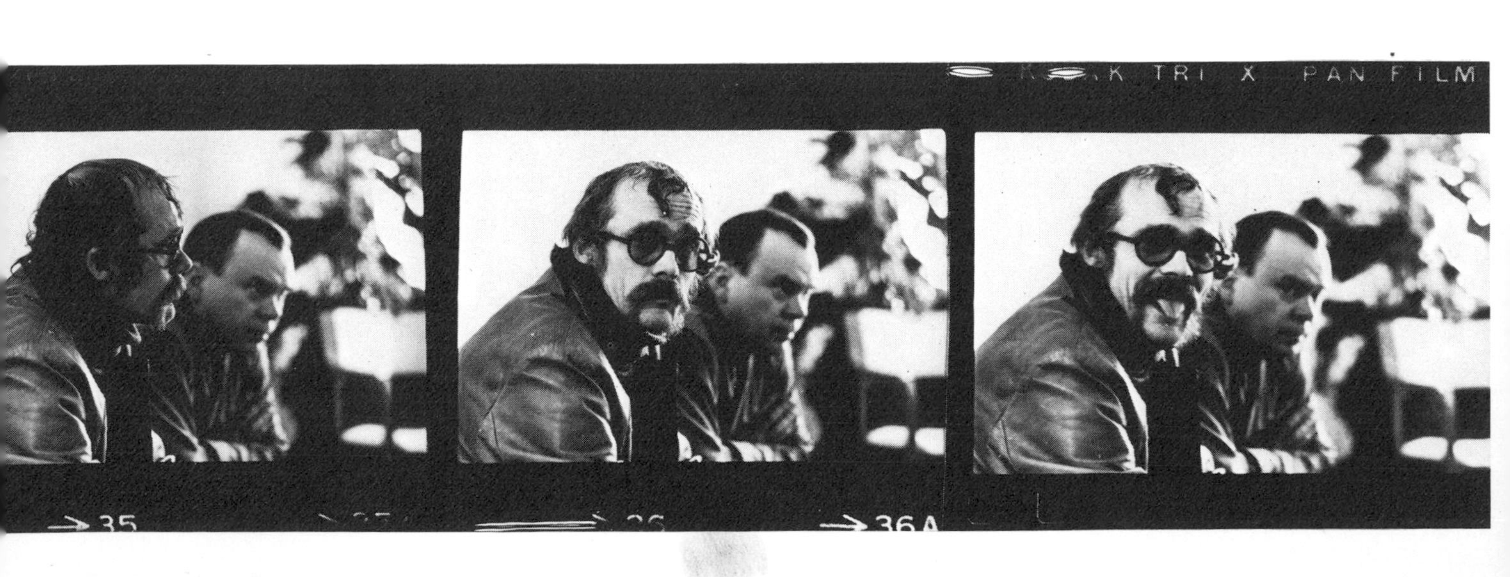

For once a book that doesn't pamper Finnish design! The author, Donald J. Willcox, is an American writer who is well acquainted with the Finnish scene. He casts a fresh look at Finnish design, a subject of nearly legendary status, alongside Sibelius and the sauna.

According to Willcox, the word »design» is a hand-me-down, a weary prostitute. It encourages myth-making, not realistic appraisal. Willcox prefers to talk of objects — and not only of them. He examines the entire field of Finnish crafts, the surroundings and conditions where this cultural form of worldwide fame is created and designed. Whatever it is — toy, furniture, clothing, or anything else — he asks: Is it necessary? Does it work? Can it work better? Is it quality? Is it reasonable? Willcox presents his evaluations in a way that is pointed, relaxed and fun. His criticism is wellmeant and constructive.

Donald J. Willcox has written extensively on craft subjects. He wrote Rya Knotting and a series of five books on contemporary design in ceramics, jewelry, stitchery, weaving, and wood. He has traveled through Scandinavia to gather material for his books and currently lives in Denmark.

Ove Hector, a Danish psychologist/photographer, has assisted in forming the abundant picture material. Household gadgets are examined, and furniture, wrapping, advertising, even problems in social planning. All the objects have been photographed in their actual surroundings, very often in a series of action shots. It is surprising how often common-sense design can be found lurking in places where one would least expect it.

VAN NOSTRAND REINHOLD COMPANY
New York Cincinnati Toronto London Melbourne